The Nocturnal City

Night is a foundational element of human and animal life on earth, but its interaction with the social world has undergone significant transformations during the era of globalization. As the economic activity of the 'daytime' city has advanced into the night, other uses of the night as a time for play, for sleep or for escaping oppression have come increasingly under threat.

This book looks at the relationship between night and society in contemporary cities. It identifies that while theories of 'planetary urbanization' have traced the spatial spread of urban forms, the temporal expansion of urban capitalism has been less well mapped. It argues that, as a key part of planetary being, understanding what goes on at night in cities can add nuance to debates on planetary urbanization.

A series of practices and spaces that we encounter in the night-time city are explored. These include: the maintenance and repair of infrastructure; the aesthetics of the urban night; nightlife and the night-time economy; the home at night; and the ecologies of the urban night. Taking these forward the book will ask whether the night can reveal some of the boundaries to what we call 'the urban' in a world of cities, and will call for a revitalized and enhanced 'nightology' to study these limits.

Robert Shaw is a lecturer in geography at Newcastle University, UK. He received his PhD from Durham University in 2012, and he subsequently worked at Durham until 2015, before joining Newcastle. With his research interests in the urban night, he has explored the production of night-time city life in the UK, changing street-lighting technologies and most recently the Nuit Debout protest movement in Paris. His work has been published in several academic journals. His personal webpage is www.rob-shaw.net, and he can be found on Twitter as @WhatIsRobShaw.

Routledge Research in Culture, Space and Identity

Series editor: Dr. Jon Anderson, School of Planning and Geography, Cardiff University, UK

The *Routledge Research in Culture, Space and Identity Series* offers a forum for original and innovative research within cultural geography and connected fields. Titles within the series are empirically and theoretically informed and explore a range of dynamic and captivating topics. This series provides a forum for cutting edge research and new theoretical perspectives that reflect the wealth of research currently being undertaken. This series is aimed at upper-level undergraduates, research students and academics, appealing to geographers as well as the broader social sciences, arts and humanities.

For a full list of titles in this series, please visit www.routledge.com/Routledge-Research-in-Culture-Space-and-Identity/book-series/CSI

Arts in Place: The Arts, the Urban and Social Practice
Cara Courage

Explorations in Place Attachment
Jeffrey S. Smith

Geographies of Digital Culture
Edited by Tilo Felgenhauer and Karsten Gäbler

The Nocturnal City
Robert Shaw

Geographies of Making, Craft and Creativity
Edited by Laura Price and Harriet Hawkins

Spaces of Spirituality
Edited by Nadia Bartolini, Sara MacKian and Steve Pile

Affected Labour in a Café Culture: The Atmospheres and Economics of 'Hip' Melbourne
Alexia Cameron

The Nocturnal City

Robert Shaw

LONDON AND NEW YORK

First published 2018
by Routledge
2 Park Square, Milton Park, Abingdon, Oxon OX14 4RN

and by Routledge
605 Third Avenue, New York, NY 10017

First issued in paperback 2021

Routledge is an imprint of the Taylor & Francis Group, an informa business

© 2018 Robert Shaw

The right of Robert Shaw to be identified as author of this work has been asserted by him in accordance with sections 77 and 78 of the Copyright, Designs and Patents Act 1988.

All rights reserved. No part of this book may be reprinted or reproduced or utilised in any form or by any electronic, mechanical, or other means, now known or hereafter invented, including photocopying and recording, or in any information storage or retrieval system, without permission in writing from the publishers.

Trademark notice: Product or corporate names may be trademarks or registered trademarks, and are used only for identification and explanation without intent to infringe.

Publisher's Note
The publisher has gone to great lengths to ensure the quality of this reprint but points out that some imperfections in the original copies may be apparent.

British Library Cataloguing in Publication Data
A catalogue record for this book is available from the British Library

Library of Congress Cataloging in Publication Data
A catalog record for this book has been requested

ISBN 13: 978-0-367-89481-8 (pbk)
ISBN 13: 978-1-138-67640-4 (hbk)

Typeset in Times New Roman
by Taylor & Francis Books

For Jack and Audrey. Still inspiring me every day.

Contents

Acknowledgements   ix

Introduction   1

1 **Changing spaces, changing times: urban futures**   7
   How is the world urban? Contemporary urban theory 10
   Urban machines: environment, society and the self 16
   Night and the limits of the city 20

2 **Fragmenting frontier: night, time and the city**   26
   Night in pre-industrial societies 27
   Electrification 29
   Production at night 31
   Night as frontier 33
   People who go bump in the night: the nocturnals 38
   Fragmentation and affects of incessancy on the edge of the city 43

3 **Nocturnal ecologies and infrastructures**   50
   Artificial lighting as a condition of nocturnal possibility 52
   Light pollution, carbon use and the ecological impacts of the night-time city 54
   Changes to urban lighting and global challenges 57
   Beyond lighting: infrastructural frontiers at night 62
   Living with darkness: infrastructure and the possibilities of urban expansion 64

4 **Nightlife and night-time economy**   68
   Booze, bingeing and beer: the night-time economy in the UK 68
   Creating nightlife: the emergence of the British night-time economy 70
   Being nocturnal: night-time subjectivities 73

Global nocturnal leisure 76
Contact zones of nightlife 79

5 **Aesthetics of the night-time city** 83
   Illuminations and conviviality: developing a night-time aesthetics 84
   Walking, exploration, graffiti: a counter-aesthetics of the night? 87
   Cities at night and place marketing 90
   The aesthetics of the night and what it means to be urban 93

6 **The domestic night** 97
   Shaping the domestic night: from comfort to control 98
   Home, subjectivity and night 100
   Home: at the edge of urban and beyond 105

7 **Towards nightology and the temporal limits to urbanism** 110
   Planetary urbanization: many cities, many planets 111
   Three narratives of the urban–world–night relationship 112
   Towards nightology 117

   Index 122

Acknowledgements

This book has emerged slowly over the last ten years, starting with my PhD research but evolving with subsequent teaching, research, conversations and publications. Inevitably, more people have helped me than I have room to name here and I don't want to list too many names: so simply, thank you to the many colleagues at Durham and Newcastle who have supported my work.

Still, a few names. It is worth highlighting the role of Paul Harrison and Gordon MacLeod in providing me with space and time to develop my ideas as PhD supervisors. While many colleagues have helped me develop my ideas over the years, Divya Tolia-Kelly, Nick Rush-Cooper and Ben Anderson have in particular provided support and advice. Tim Edensor's time in Durham during 2014 was very important in helping me expand the scope of my research interests beyond the night-time economy, and collaborations or conversations with Casper Ebbensgaard, Ankit Kumar and Gerry Taylor-Aiken have helped me at different moments along the way. At Newcastle, my colleagues have been more welcoming than I could imagine, but Jon Pugh, Michael Richardson and Helen Jarvis have all been particularly supportive.

Then of course there is Jenny, who first as my colleague, then as my friend and eventually as my wife has helped me both intellectually and personally in ways that I could not even begin to thank her for. Lucy entered our lives when this book was nearly finished, and night-time feeds have been an appropriate time for me to ruminate over the final edits. My parents, sister and wider family have always been supportive too; I consider myself very lucky.

A number of people have looked at chapters or extracts of this book during the writing process. Some of them are mentioned above; in addition, Alastair Bonnett, Colin McFarlane, Lizzie Richardson, James Ash, Jon Silver, Gareth Powells and Matt Jenkins have all looked at draft versions of parts of the text. Any errors or limitations are clearly mine, not theirs.

Introduction

To enjoy this book at its best, you might want to create your own experience of night. If it's light outside, you'll need take a series of actions. Draw your blinds and turn off any overhead lights. Reading lamps or table lamps will do just fine. Dim your computer screen a little if it's on – there's no need for that bright electronic glow here. Maybe make yourself a drink – a hot drink if you want, or something a little stronger if you feel like it – and a snack if you're committing to reading this for a while. Snuggle down, and revel in your mini-experience of night, even if it is midday outside. The night that you're in right now, however, could be ruptured at any moment. A telephone call; a parcel delivery; a friend or colleague knocking on your door. These would not happen in the night, or at least would be highly unlikely, but all of them would disrupt the nocturnal atmosphere that you have created. It is a fragile imitation of night, therefore, a faint replication of the real thing. Night cannot be reduced to one essential feature. It is not darkness. It is not sleep. It is not the time in which services close down. Rather, night is multiple. It is atmospheric, it is affective, it is subjective, it is natural, it is social, it is static, it is rhythmic. While we can create nocturnal scenes or features, we can neither expect to be able to produce 'night' easily, nor can we expect the night to be untouched by interruptions and perturbations.

This book is an attempt to put the night at the centre of geographical enquiries into the city and the world. The urban night is a time of fascination for many, in which people enjoy a drink, explore the city's dark corners, view the planets and stars, or go in search of a life outside of the surveillance that the light of day brings. Night appears in a range of research areas, from exploration of women's safety in accessing public space through to studies of infrastructure maintenance. However, night is rarely the focus of research in and of itself. As a 'natural' phenomenon, the night is often overlooked: a forgotten background with which we are all familiar and which as such does not present itself as an obvious topic for research. Furthermore, the majority of our research is diurnal. Ethnographers visit organizations during the day, and then go home. Interviewees are asked about their daily rhythms. Surveys and questionnaires speak of days, not nights. Even where night comes up, nocturnality is rarely taken as the topic of research. Researchers looking at 'the night-time economy', for example, focus almost universally on the presence of alcohol and the governance of its consumption. If temporality appears as a research topic in this area, it is typically in relation to the rhythm and scheduling of bar opening hours and licensing laws. Meanwhile, research which does explore temporalities is often focused on phenomenological experiences or rhythms, as opposed to constructed periods of time such as night. That something such as the night-time economy is the *night-time* economy is barely explored: the difference that night makes to the alcohol and leisure industry is largely absent.

Nonetheless, a growing range of researchers are exploring topics which relate to the night. The social life of sleep and rest has emerged as an interesting interdisciplinary field between science, the humanities and social science (Callard, 2014). How we inhabit the night shapes how we inhabit day, as society moves towards rhythms of 24/7 (Crary, 2013). Research in this area has repeatedly reinforced the value of sleep and rest, despite the ways in which it is attacked. Another field of blossoming research has been the exploration of experiences, constructions and productions of dark and light, across multiple disciplines (Dunn, 2016; Edensor, 2017; Jóhannesson and Lund, 2017; Pritchard, 2017; Stone, 2017). As explored in this book, light and dark are understood as relational, coming as a pair rather than as separate entities. Rarely, if ever, do we actually experience pure dark or light; rather, we encounter a world of shadows, of spotlights, of overcast skies, of glowing firelight and so on. It is these shifting patterns of light and dark that shape our world at both day and night. Noted previously, 'night-time economy' research has driven the nocturnal agenda in much of social science, exploring the ways in which cities are opened up at night for the alcohol and leisure industry. Researchers have developed a diverse picture as to the range of drinkers and drinking practices, attempting to show how night-time economies globally are more than binge drinking (Jayne *et al.*, 2011; Eder and Öz, 2014; Fjær *et al.*, 2016). Studies which go beyond this one element of the night-time economy are rare, however. Other night-time activities that have had some attention have been routines of repair and maintenance (Graham and Thrift, 2007; Carr, 2017), in which the role of night as a repair time for both infrastructure and broader social life is emphasized. Night shift work has been a major focus of health and psychological research (Wang *et al.*, 2014), but much less has been done to explore how this intersects with wider themes in social science. So night and related topics are beginning to appear in the literature, but rarely with nocturnality as a focus.

A key aim of this book is to put the night front and centre in the research agenda. How do we understand (some) of these issues if we make the night itself the question that we are exploring? Night offers an interesting lens because of the ways in which it straddles the social and the natural. Night is both an inherent feature of earthliness, bound up in the evolutionary history of all animals on the planet, and a socially produced feature of routines and rhythms. The natural and social elements cannot be untangled; they work in unison, though as residents of high latitudes in mid-winter will tell you, not always in harmony. Night is similar to what Deleuze and Guattari would call a 'haecceity': an assemblage which consists 'entirely of relations of movement and rest between molecules or particles, capacities to affect and be affected' (Deleuze and Guattari, 1987, pp. 287–288). In other words, night has a being different to a thing, a subject, a person, an animal; it is a collection of relations, possibilities and materialities. While there are material features to night, and elements of night which act in unison, there is no material, thing or form that we can point to as 'night'. I cannot hold a lump of night in my hand, but I can take you to a city and show you its night. So night is much more of a collection of precarious relations than any one thing or set of things. To close down the breadth of this haecceity, this book focuses on night in cities, drawing on this timespace because of the ways in which the mixing of the planetary and the social that defines night speaks to contemporary debates in urban geography. Specifically, as urban forms and infrastructure have spread and urbanization continues across the globe, researchers have asserted that we are entering or have entered an area of 'planetary urbanization', in which the whole earth becomes subject to the conditions of urban life. The strongest proponents of this theory see this urbanization as a continuation of the expansion of global capital, as ever more of the globe becomes part of a single urban system (Merrifield, 2013; Brenner and Schmid, 2015). This argument stretches traditional claims about urban hinterlands to argue that the spread of capitalist relations, infrastructure and the capacity to live an urban life in the most 'rural' of areas mean that the city can now be said to be at one with the planet.

In contrast to this Marxist-rooted argument, proponents of what Derickson (2015) calls 'planetary urbanization 2s' offer a similar but perhaps more cautious description of planetary urbanization drawing from postrustrucalist and postcolonialist theory. These theories see the contemporary urban moment as characterized by increased exchange between cities, with urban conditions in Global North and Global South becoming more and more alike, and the urban form increasing in dominance. While such theories argue that the world is highly urbanized, they are less likely to claim that this urbanization is, or will be, total. Instead, they emphasize the unequal and differentiated nature of urban living, looking for comparisons between the multiple different forms of city life across the globe.

Night in the city can interject into this debate by bringing together planetary being and socialized being. To inhabit the urban night is to inhabit both the earth and the constructed city environment. As such, it offers an interesting lens for exploring these arguments about the planetary nature of urban life in the twenty-first century. In particular, the night points to the idea that being 'planetary' is more than being 'global'. Being planetary involves experiencing the conditions of living on earth as a planet. Night is one such condition; it is inevitable, even if developments in technology mean that it has been dramatically altered. In this book, I argue that if we understand being as planetary, then we need to follow the suggestions of Naess (1989) and Guattari (2000) and develop an ecosophical understanding of the earth. In other words, we need to put the knowledge of planetary being and our relationship to earth as a home (the Greek *oikos*, which has resulted in our contemporary prefix 'eco-', meaning both 'home' and 'family') at the heart of our research. In addition to this, introducing the night as a topic of enquiry brings in questions about the rhythms of the city in a planetary context. In other words, are there spaces which are 'urbanized' at day but not urbanized at night? Would it be meaningful to continue to speak of temporal boundaries to urbanism in the world, even if spatial boundaries are being or have been eroded? Exploring some of these questions, and understanding the difference that night makes, is the key aim of this book.

The Nocturnal City draws from my own research experience, from the wider published literature and from social and geographical theory. Inevitably, there will be some gaps. The book could say more about the themes of rest and relaxation, and it does not explore the research on night shift work that to date has mainly focused on physical and psychological health. The organisation of the book means that issues of race, gender, sexuality and other identity positions are spread throughout the work, rather than forming a central part of any one chapter. In part, this is a result of the argument that the urban night is a 'subjectivity machine', producing both urban space and the people within it; the book focuses more on these underlying processes and their mechanisms, rather than the experiential components of the urban night that emerge from them. Nonetheless, it is worth highlighting here that the subjects produced in the night-time city do not encounter it equally. Night-time cities are spaces in which, broadly speaking, men are more freely able to choose to enter, while women are more likely to be forced to enter for work. People from ethnic minorities in any given country are more likely to suffer violence at night than at day. While historically used as cover from surveillance by LGBT people or political dissidents in many countries, the night also has dangers in it for both groups. The reduction in visibility and increase in conviviality that come with night do not remove the structured problems of day.

Chapters 1 and 2 provide conceptual underpinnings for the book. In Chapter 1, I position the book within urban theory, and in particular the debates surrounding planetary urbanization that I have introduced here. By bringing the writing of Felix Guattari to the forefront in particular, I hope to offer fresh perspectives on the urban–world relationship that is at the heart of the planetary urbanization debate yet perhaps somewhat overlooked. Guattari's theorization of the 'Three Ecologies' of self–society–earth provides an especially useful framework for considering what is brought together in the urban night, and how we might continue to conceptualize *planetary* urbanization. Chapter 2 turns to conceptualizations of the night in social science, with a view to bringing out how it has been understood as an object of research. Central to this chapter is the argument that night should be understood as a 'fragmenting frontier', a borderland between urban and non-urban that is becoming increasingly stretched.

Chapters 3, 4 and 5 all illustrate some key fields in which researchers have explored the night. Chapter 3 focuses on infrastructures, using lighting as its key example; it introduces the concept of the 'biogeoastronomical' night to describe those elements of night which emerge because of planetary living. Chapter 4 discusses the night-time economy and attempts to connect the best-researched case – the UK – to global night-time leisure in cities. Chapter 5 is a discussion of the aesthetics of the night that have emerged, contrasting the city as spectacular with the city as spectacle: that is, the urban night as a site of display versus the urban night as a site for exploration. Here, I focus on night as aesthetic to try to get at its use as a way of promoting ways of living in the city.

Chapter 6 moves us beyond the public sphere and into the night at home. It looks at how the domestic night troubles narratives of planetary urbanism as a way of asking whether we can find in the night-time city some of the boundaries to this idea, the outside of the city that Brenner (2014) claims does not exist.

In Chapter 7, I conclude with a reflection on these temporal limits to the city, and offer a few suggestions as to what a future 'nightology' might study.

The urban night is a fruitful place, both dark underbelly of the city and a core element of how we must understand urbanism as a way of life. This book brings together a range of ideas from many fields and explains how these can help us unpack the urban night, and in so doing understand planetary urban life. The night has appeared in fragments and moments in a range of different research projects. Taken as a whole, it provides a new opportunity to consider the conditions of planetary urban living – the ways in which earth, society and self come together uniquely in the built environment.

References

Brenner, N. (2014) 'Introduction: Urban theory without an outside', in Brenner, N. (ed.) *Implosions/Explosions*. Berlin: Jovis, pp. 14–31.
Brenner, N. and Schmid, C. (2015) 'Towards a new epistemology of the urban?', *City*, 19(2–3), pp. 151–182.
Callard, F. (2014) 'Hubbub: Troubling rest through experimental entanglements', *The Lancet*, 384(9957), p. 1839.
Carr, C. (2017) 'Maintenance and repair beyond the perimeter of the plant: Linking industrial labour and the home', *Transactions of the Institute of British Geographers*, online early access: http://onlinelibrary.wiley.com/doi/10.1111/tran.12183/full (accessed 17 October 2017).
Crary, J. (2013) *24/7*. Los Angeles: Verso.
Deleuze, G. and Guattari, F. (1987) *A Thousand Plateaus: Capitalism and Schizophrenia*. Minneapolis: University of Minnesota Press.
Derickson, K.D. (2015) 'Urban geography I: Locating urban theory in the "urban age"', *Progress in Human Geography*, 39(5), pp. 647–657.

Dunn, N. (2016) *Dark Matters*. Arlesford: Zero Books.
Edensor, T. (2017) *From Light to Dark*. Minneappolis: University of Minnesota Press.
Eder, M. and Öz, Ö. (2014) 'Neoliberalization of Istanbul's nightlife: Beer or champagne?', *International Journal of Urban and Regional Research*, 39(2), pp. 284–304.
Fjær, E.G., Pedersen, W. and Sandberg, S. (2016) 'Party on wheels: Mobile party spaces in the Norwegian high school graduation celebration', *British Journal of Sociology*, 67(2), pp. 328–347.
Graham, S. and Thrift, N. (2007) 'Out of Order: Understanding Repair and Maintenance', *Theory, Culture and Society*, 24(3), pp. 1–25.
Guattari, F. (2000) *The Three Ecologies*. London and New Brunswick, NJ: Athlone Press.
Jayne, M., Valentine, G. and Holloway, S.L. (2011) *Alcohol, Drinking, Drunkenness: (Dis)orderly Spaces*. Aldershot: Ashgate.
Jóhannesson, G.T. and Lund, K.A. (2017) 'Aurora Borealis: Choreographies of darkness and light', *Annals of Tourism Research*, 63, pp. 183–190.
Merrifield, A. (2013) 'The urban question under planetary urbanization', *International Journal of Urban and Regional Research*, 37(3), pp. 909–922.
Naess, A. (1989) *Ecology, Community and Lifestyle*. Translated by Rothenberg, D. Cambridge: Cambridge Unviersity Press.
Pritchard, S.B. (2017) 'The trouble with darkness: NASA's Suomi satellite images of earth at night', *Environmental History*, online early access: https://academic.oup.com/envhis/article-abstract/22/2/312/2998686/The-Trouble-with-Darkness-NASAs-Suomi-Satellite (accessed 17 October 2017).
Stone, T. (2017) 'The value of darkness: A moral framework for urban nighttime lighting', *Science and Engineering Ethics*, online early access: https://link.springer.com/article/10.1007/s11948-017-9924-0 (accessed 17 October 2017).
Wang, F., Zhang, L., Zhang, Y., Zhang, B., He, Y., Xie, S., Li, M., Miao, X., Chan, E.Y.Y. and Tang, J.L. (2014) 'Meta-analysis on night shift work and risk of metabolic syndrome', *Obesity Reviews*, 15(9), pp. 709–720.

1 Changing spaces, changing times
Urban futures

On New Year's Eve, around the world, people in cities, towns and villages celebrate in a routine choreographed according to the globe's complex and geopolitically constructed time zones. Where and when the New Year begins is somewhat contested: in 1995, Kiribati shifted the time zone of its most easterly islands, ostensibly to iron out logistical difficulties of having land either side of the International Date Line, but conveniently making Caroline Island the first inhabited land to reach the year 2000, an event which resulted in the island being renamed 'Millennium Island' and hosting a celebration viewed by a billion people across the world. Starting in Kiribati, New Year celebrations travel roughly westwards, bringing to cities in the middle of the night fireworks, public gatherings, dances, concerts, public broadcasts and other celebrations. Twenty-six hours later on Baker Island – actually a few hundred miles to the northwest of Millennium Island – the final celebrations are held by the few US naval personnel serving at this isolated base.

Between Kiribati and the USA, then, this celebration incorporates local, regional and national cultures, at the same time as marking a globalized system in which a European–Christian calendar has become the international standard. This nocturnal celebration is at once simultaneous – in that it occurs in all countries at midnight – and separated, travelling around the world. It is made up of both small moments (house parties, individuals with fireworks, families seated around the television) and large moments (spectacular events, with cities such as Sydney, Dubai, London, Paris and New York competing with one another to host the biggest or most extravagant firework displays). It is marked by its mundanity, as an annual reoccurrence, in which many people will not participate, but also by its rupture, as a moment of celebration and unity into which many people put a lot of emotional energy. It is notable in the almost unique centrality it places on mid*night*: while other festivals or celebrations contain nocturnal components, no other event brings a worldwide focus to the night at the same time, across cultures.

This moment and the contradictions that it contains reveal many of the difficulties that we face when considering how a globalized world connects together. How do we resolve the gaps between the specific or the individual and the general or the collective? Cities are sites in which many people collectively experience the same events, emotions, routines and so forth, but in which these experiences are vastly differentiated according to both individual circumstance and large socio-cultural groupings. As many people are alienated and excluded from urban life as are enchanted and enraptured by it. We could if we wanted describe New Year's Eve in terms of the flows that go into making 'event capitalism'; we could write with reference to the society of the spectacle, or to Baudrillard's simulacra; we could contrast the excess of celebration with the waste and damage that it creates; we could discuss it as an event of globalization; we could explore the role of ritual within society; we could take a practice theory or actor-network approach, showing the complex processes of organization and coordination through which such events occur. To a greater or lesser degree, all these approaches have validity. However, each of them would tell us only part of the story about New Year's Eve. They would reveal some of the ways in which different relations result in the production of spaces, moments and events in which the very being of urban dwellers is created out of their relations to the bodies and events around them. However, each narrative would also hide something, elide differences and overlook complexities.

In this chapter, I will offer my own narrative in relation to global urbanism. The aim will be to show how both conceptual development and empirical exploration of the urban night can expand our understanding of global urban life more broadly. I do so with the aim that this narrative and theorization be understood as adding something to previous accounts, rather than correcting or contradicting them. Therefore, I will address *the city* in this chapter before providing an introduction to *the nocturnal* in Chapter 2.

My approach begins with the premise that cities are machines which through all of the processes described above produce the people who inhabit the city, the built environment and the social norms that define urban life. My reading is inspired by the French philosopher, psychologist and activist Felix Guattari's conceptualization of the machine:

> If one broadens the concept of machine beyond its technical aspects and takes into account its economic, ecological and abstract dimensions and even the desiring machines that people our unconscious drives, one must treat the mass/aggregate of urban and architectural machinery as machinic components, all the way down to their smallest subgroupings.
>
> (Guattari, 1993, p. 145)

Guattari is not the typical first choice of the urban theorist, but his work offers an interesting way for thinking about cities. In particular, I find him useful in helping frame urban life as at the intersection of what he calls the 'environmental ecology, social ecology and mental ecology' (Guattari, 1993). In other words, urban life emerges jointly out of the natural/built environment, social norms and power relations, and the psycho-social sense of self and identity. In a 1990 public lecture titled 'Space and Corporeity', given just eighteen months before his death, he predicted that urban studies would face the need to respond to shifts in the relationship between these ecologies and, as he outlined in more detail elsewhere, the necessary response would need to be an 'ecosophy' (a 'knowledge of the home') that would involve producing an 'ethico-aesthetic paradigm' (Guattari, 1995; 2000). In other words, the challenge for future urban studies would be to find ways of producing understandings of cities as inhabited machines which produce. In developing this understanding, researchers might offer small insights into the production of more egalitarian societies in which people live with both the 'natural' and 'built' environment around them. Guattari, in his own writing and in his collaborations with Gilles Deleuze, developed a vision of the social world as emerging out of the intersections of the environmental, social and psychological. In conversation with contemporary urban theory, this can provide what is for me a persuasive vision of the urban world that can help us answer questions about changes to urban life today.

As cities become bigger, as speeds of communication become faster, as technology becomes more pervasive, the urban subjectivity machine is changing in both positive and negative ways. To focus on the more negative challenges that this transformation of urban life is bringing is to focus on multiple new and emerging challenges. Extreme rural poverty is being replaced with precarious urban poverty. Droughts and famines may become rarer, but insecurity increases in the informal settlements of megacities around the globe, in the deindustrialized cities of many developed counties and in the cities globally that experience a daily threat of terrorism. Basic services, information and other facilities are now more readily accessible by greater percentages of the globe's population than ever before, but at the cost of reduced interactions with people, increased centralization of power, and a series of increasingly negative and alienating affective experiences (Berlant, 2011). Urban violence replaces rural oppression. Gentrification dispossesses the poor, so that the richest cities become more and more the playthings of a small, elite global group. In urban studies, the current desire to rethink the city can be connected to the empirical and conceptual recognition of these trends, which were predicted by many key theorists of the late twentieth century.

But in this list of contemporary problems, why then look at the urban night? My answer is that the night is uniquely placed at the intersection of the three ecologies (environment, society and self) that Guattari identifies. It helps us understand the intersection of these three fields, and the way in which they mechanically produce urban life. So while this book is clearly not the place for answering all the challenges of the contemporary city, it can offer two key contributions. First, it uses the urban night as a useful case study: how has this particular timespace been transformed, altered and shaped as cities have grown and time has accelerated? Almost uniquely, 'night' is a feature of every single city across the globe; some version of it is universal. As we shall see, during the twentieth century the night was repeatedly characterized as a time that was being 'colonized' by day; a 'frontier', akin to the spatial frontiers of the colonial era, into which daytime practices were expanding (Melbin, 1987; Schivelbusch, 1988; Gwiazdzinski, 2005). While globalization might mark the point at which this temporal expansion is completed, the urban night (like spatial frontiers) has continued to be a fluid and changing space, and this book aims to unpack some of these transformations.

Second, night can help us comment on the contemporary limits of urban life. If postcolonial work has shown how this depiction of the spatial frontier was at best a partial one which is becoming increasingly irrelevant (Roy, 2011), perhaps we should pay similar critical attention to this 'temporal frontier'. Extending the arguments further, this focus on the intersecting spatio-temporal boundaries can inform some of the debates on the nature of the city that are emerging through these transformations.

Before turning to the night, however, I want to use this chapter to explore this understanding of the city in more detail. Urban studies has, in the 2010s, been an area of intensive and extensive theorization, in which there have been several attempts to use different frameworks to conceptualize urban life.

How is the world urban? Contemporary urban theory

> The degree to which the contemporary world may be said to be 'urban' is not fully or accurately measured by the proportion of the total population living in cities.
>
> (Wirth, 1938, p. 2)

Wirth's simple insight is a fundamental moment in urban studies. From this point onwards, the relationship between the built environment and 'the urban' starts to separate. The concept of the 'urban' or of 'urban life' becomes associated with a series of characteristics that extend beyond the presence of a dense population in an extensive built-up environment. Diversity of population, access to services, transport connections, presence of cultural and other activities, intersections of networks of business, manufacture, trade and finance, intensity of daily life and other concepts all become associated with 'the urban'. Through the twentieth century and into the twenty-first, urban theorists have continued to grapple with this disconnection in order to try to explore the processes and practices through which urban life is produced, and the broader socio-spatial impacts that this urban life has. Urban studies has moved through a series of different and intersecting understandings of how the world's cities relate to one another, and the daily life of people in cities or living urban lives. Throughout this time, a series of 'minor trends' has sought to correct, push or challenge the dominant theorizations of any one moment. Ball argues that 'the urban' in many theories might be understood as either a 'pragmatic concept with which to deal with empirical material or one spatial level out of a series of highly interlinked ones' (Ball, 1986, p. 448). In other words, regardless of the particular focus, the city has typically been understood as either a series of empirical characteristics (as with Wirth) or as a particular scale within a hierarchy or series of scales. Throughout this history of theorization and retheorization, the question of what the city is or how cities connect to one another has thus come and gone.

It is fair to say that, at the time of writing, we are at a more intensive point of questioning the nature of cities. There is clearly no set starting point for the contemporary debate, but we can identify a series of prompts, both empirical and within the world of academia. As Hall and Savage (2016, p.82) summarize:

> [T]he accelerated expansion of the urban in the landscapes and mind-sets of the twenty-first century has been accompanied by a renewed interest in comprehending current processes of worldwide urbanization. Longstanding questions of definition – 'what is the city?'; 'what is the urban?' – are posed with new urgency as we engage with urban dynamism across the planet.

We might call this new worldwide urbanization the era of 'post-globalization' to refer to a planet in which the period of globalization has now happened. In this era, the 'West versus the rest' hierarchy established during the nineteenth and twentieth centuries has not quite collapsed but it is becoming frayed, with flows of people, material and information moving from city to city across the globe. What were once dominant states or businesses in the Global North are now met by cities, corporations and states from nations of the Global South; furthermore, these two categories are becoming increasingly less useful. Such moves have been associated with both economic growth outside of traditional power centres and stagnation in Europe and North America. What this is not, however, is a period in which North and South have become matched in wealth and opportunity. Rather, there has been a simultaneous global incorporation of more and more spaces into a single socio-economic system alongside increased fragmentation within that system, bringing Smith's (1982) concepts of equalization and differentiation to mind. Wealth has been increasingly concentrated among elites who are free to travel and take advantage of global mobility, with the poor in all societies dropping back. Thus, the elite 'world cities', whether defined loosely or according to a series of metrics (Smith and Doel, 2011), have become increasingly dominant over the spaces around them. Furthermore, urban growth has been notable in its variation. It is well recognized that the cities of East Asia and the Middle East have grown by multiple orders of magnitude in under fifty years. Perhaps even more remarkably, older shrinking cities have in many locations seen new growth after decades of urban decline: for example, between 2001 and 2011, Manchester in the UK experienced its first population growth for forty years, and its fastest growth for a hundred years. Famously, various international organizations declared various years in the late 2000s as the year in which the majority of the globe's population became urban, according to the different metrics that they used. As well as the sheer number of people living in cities, the emergent 'megacities' present a series of new phenomena, challenges and features for understanding cities.

So the academic desire to rethink and discuss the city has come from this repositioning of cities in global systems, and new forms of urban living. We can identify a step towards questions about the nature of cities occurring in the mid-2000s with Mike Davis's *Planet of Slums* and Jennifer Robinson's *Ordinary Cities* both offering positions that seem to mark the emerging divisions in understandings of cities (Davis, 2007; Robinson, 2006). Davis argues that the model of 'world cities', with its origins in Marxist 'core–periphery' models of global urban economics (Friedmann, 1986), is coming to an end. He sees the point of transfer from a majority-rural to majority-urban world as the moment when urban life becomes characterized by inhabitation not of cities as we had previously recognized them, but of either large megacities stretching over vast regions or 'cityized' suburbs, towns and villages which lose their rural characteristics to become effectively urban: 'in many cases, rural people no longer have to migrate to the city: it migrates to them' (Davis, 2007, p. 9). The result is a changed dynamic to the globe's cities, in which 'the exploding cities of the developing world are also weaving extraordinary new urban networks, corridors, and hierarchies' (p. 5). Davis's analysis pulls apart the dystopia of contemporary capitalism, arguing from a perspective infused by Marxism and postmodernism that slums represent a future for cities in a world in which surplus capital has been translated into 'surplus humanity'. He has little time for what he calls 'portentous post-Marxist speculations' (p. 201) about multiplicity and multitude.

As such, Davis's analysis contrasts starkly with that of Robinson in *Ordinary Cities*. She sees urban studies as having analysed cities according to a binary of 'modern' and 'not-modern' in which two types have been studied: wealthy, global, Western cities as examples of the modern, and poor megacities as examples of the not-modern. Her critique is twofold. First, she aims to 'dislocate modernity' in order to move beyond a singular vision of modernity so that we start to understand the modern as consisting of 'many different cultures in many different places ... [which are] enchanted by the production and circulation of novelty, innovation and new fashions' (Robinson, 2006, p. 7) The vision here is of a modernity which is multiple, which contains within it innovation, changes and difference produced from across the globe. Robinson's suggestion is that urban studies needs to open itself up to ideas and comparisons from a wider range of sites, particularly those outside the Global North, in order to become 'more cosmopolitan' (p. 168). The second part of her critique is that urban studies needs to source its ideas from a wider range of cities in all contexts. This is the meaning of her term 'ordinary cities', which 'draws our attention to cities as distinctive assemblages of many kinds of activities ... [with] the capacity to shape their own futures, even if they exist in a world of (power-laden) connections and circulations' (p. 170). Urban studies, for Robinson, needs to reject a focus solely on the largest and most powerful cities and instead use comparative methodologies to explore a diverse range of urban experiences. One might add that this means researching not just the experiences of people in city centres but also those of suburbanites, or those who are marginal in cities, or those who inhabit cities only transiently and so forth.

While differing in much of their analysis, both Robinson and Davis offer visions, to some extent at least, of an 'urban world': for instance, '[urban studies] needs to be able to be attentive to the diverse experiences of a world of cities' (Robinson, 2006, p. 94); and 'the earth has urbanized even faster than originally predicted' (Davis, 2007, p. 1). In other words, the need to explore the nature of cities, how they relate to one another and how we understand them comes at the moment of a 'post-globalization' vision of the world in which cities no longer appear as a discrete scale among other scales, but instead become major actors or assemblages of actors on a planetary basis. Cities gain as part of the shift away from scalar socio-economic hierarchies, which developed in previous eras of imperialism and capitalism, and towards a more diverse, heterogeneous system in which wealth and power are located at nodes across networks, spread more widely and thinly. In a global context, these changes have reshaped relations both between and within cities.

Robinson's and Davis's books mark the start of a period of urban studies attempting to deal with these challenges – reminiscent of the challenge that Guattari predicted would be coming to the discipline (Guattari, 1993). The spread of this retheorization of the city is remarkable: a Google Scholar search for the term 'urban theory' returns 3,500 results from 1996 to 2006, and 11,300 results from 2007 to 2016. This fertile conceptualization has produced a variety of different ideas, models, theories and approaches. To position my work, I want to approach three of these, all of which lead us to a vision of the city as a subjectivity machine: first, postcolonial approaches that have developed from the work of Robinson and others; second, approaches that have built on practice-oriented work and particularly the concept of 'assemblage'; and, third, Marxian analyses which have coalesced around the concept of planetary urbanization.

In light of Robinson's work, but also responding to the wider demands of postcolonial theories, geographers have sought to add to the range of cities, concepts and voices that contribute to the debates on urban life across the globe (McFarlane, 2010; Ong, 2011; Sheppard *et al.*, 2013; Derickson, 2015; Robinson and Roy, 2015). At the core of this approach has been what (Ong, 2011, p. 12) describes as a 'mid-range theorizing', in which the aim is not to replace pre-existing metanarratives of the city with a new metanarrative, but to offer explanatory frameworks that are embedded in case studies which offer 'ambitious practices that creatively imagine and shape alternative social visions and configurations'. There are two key commitments in this area of theorization. The first is methodological, relying on scholars to embed themselves in a variety of different social positions across the globe (McFarlane, 2010; Derickson, 2015). This has been labelled a 'comparative' approach to urbanism, but the act of comparison here is imagined in a quite specific way. As McFarlane (2010, p. 726) argues, all claims about 'the city' are comparative, in that they draw from our situated knowledge or are set against some sort of other: either how cities are, or how we imagine they could be. In thinking comparatively about the city, McFarlane suggests that we need to consider comparison as a 'strategy', an ethico-political move, in which we make a commitment to learning from the conceptual and empirical realities of different cities. Robinson (2011, p. 18) argues similarly that comparison should help 'strengthen urban studies' engagement with different models of explanation and causality' as an act of 'proliferation'. The commitment towards the comparative is thus a commitment to the recognition of the specificity, complexity and performativity of different urban cases.

The second commitment implied in postcolonial urbanism has been an imagination of the new: that is, of new ways of making or doing the city. Postcolonialism has been keen to turn to diverse practices of using and shaping urban space from across the globe in order to understand potential different ways of inhabiting the city. Derickson (2015) reviews the contributions made in this area, arguing that this work is an 'urbanization 2s', a series of secondary, overlooked ways of doing and living the city that are brought into focus through careful empirical attention. What is important here is that a diverse range of urban experiences are not simply 'wheeled out' as case studies, but are created on the basis of sustained and detailed engagements between scholars and those who live in marginal city spaces (Silver, 2014).

The postcolonial approach, which has sought to diversify the way we think about the city through a series of grounded conceptualizations that bring in difference, has been broadly allied with the area of research that has drawn from a variety of trends in practice theories. Though various different labels might be applied to this work, which incorporates a variety of influences, the orientation around the concept of 'assemblage' is a consistent theme in urban studies (McCann and Ward, 2011; McFarlane, 2011a; 2011b; Farias and Bender, 2012; Derickson, 2015). Here, the globe's cities are understood as both being themselves and forming as a group into self-organizing systems which come together through the actions of a wide range of different actors. 'Assemblage' itself is a term that has been used in a variety of ways: McFarlane (2011a, p. 206) describes it as an orientation, an imaginary and a concept, but, crucially, it is 'an attempt to describe relationalities of composition ... Rather than focusing on cities as resultant formations, assemblage thinking is interested in emergence and process, and in multiple temporalities and possibilities.' As a concept, assemblage looks to explore how relations emerge from interactions, how these relations do or do not hold together, and the performative effects that these relations have. This move has encouraged urban studies towards a proliferation of sites and actors in the consideration of what makes the city. In other words, assemblage and its related concepts push us towards looking at unintended relationships, flows which ambiguously change and shape, and agency that can be dispersed into 'inanimate' matter (Bennett, 2005). In combination, postcolonial and assemblage urbanism have pushed urban studies both outwards and inwards. Outwards, these approaches encourage researchers to push towards a wide range of urban experiences from across the globe; inwards, they encourage researchers to delve into the differentiated and complex, interconnected experiences of urban life.

Standing in contrast to these two approaches is the 'planetary urbanism' description of the globe's cities, an evolution of Marxist world systems theory. As with the previous understandings of the city, this approach centres on the notion that there has been a radical change in the nature of what constitutes urban life:

> Following the crisis of national-developmentalist models of territorial development, the collapse of state socialism and the subsequent intensification of global economic integration, a variety of contradictory urban transformations has been under way. The causes, contours, contexts, interconnections and implications of such transformations are widely debated, and remain extremely confusing in the wake of the global financial and economic crises of the late 2000s and early 2010s.
> (Brenner and Schmid, 2015, p. 151)

As with postcolonialist and assemblage thinkers, there is a clear intersection of intensifying and complex global social and economic flows, and the pyscho-social experiences of living in cities. By contrast to the previous approaches, however, the planetary urbanism movement sees this as much less of a *break* from previous trends, and more a continuation or evolution of the related processes of the spread and growth of capital, and the spread and growth of the city. Of note, Brenner and colleagues locate their arguments firmly in the context of Lefebvre's claim that 'society has been completely urbanized ... this urbanization is virtual today, but will become real in future' (Lefebvre, 2003, p. 1). Lefebvre goes on to clarify that the urban society he describes is one which has undergone a 'series of discontinuous transformations [which] burst apart ... old urban forms' (p. 2). In other words, for Lefebvre, the urbanization of the planet and the transformations that have happened to cities and urban life as part of this have followed a series of stages, which can be traced back to the spread of capital. Here, it is also worth remembering that Lefebvre's wider aim is to use the concept of the dialectic to explore the ways in which capital spreads socially, temporally and spatially. Unlike many of his contemporaries, he recognized that capital spreads by embedding itself in our routines, habits and practices: that is, in our rhythms (Lefebvre, 2004). As such, Lefebvre explores how power and domination associated with capital works itself into the fabric of everyday life through a heterogeneous set of interconnected practices. In identifying the formal elements of the city – planning, governance and so forth – as key tools of capital, Lefebvre argues that urbanization is one of the key outcomes of this spread of capital.

In *Towards a New Epistemology of the Urban*, Brenner and Schmid outline how these arguments are developed under theories of planetary urbanization. First, while not rejecting all the insights of postcolonial- and assemblage-oriented theorizations, they argue for a singular *epistemology* of 'the urban', which allows for a multiplicity of urban *forms*. In other words, Brenner and

Schmid do not call for anything that looks like a 'model city' or typology of cities, but they do claim that 'in a capitalist world system ... contextual specificity is enmeshed within, and mediated through, broader configurations of capitalist uneven spatial development and geopolitical power' (Brenner and Schmid, 2015, p. 161). That is, urban forms may be unique and contextual, but they are driven by systemic connections. This system extends beyond built-up areas into peri-urban, rural and agricultural areas which are 'now increasingly internalized within world-encompassing, if deeply variegated, processes of planetary urbanization' (p. 163). The claim here is that while we can diversify and broaden our understanding of the sites which form part of 'the urban', this is an expansion of the category of the city rather than a destruction of it. They thus characterize their project as attempting to create a suitably open and diverse theory to incorporate the broad phenomenon of the urban, while not losing sight of the global dominance of a capitalist world system. Their 'epistemology of the urban' proceeds via seven points. I won't outline all of these here, but their key claims centre on Lefebvre's argument that the 'urban' is a conceptual – rather than an empirical – object; that urbanization has become increasingly pervasive and heterogeneous; that such developments are highly uneven across multiple scales; but that all of this nonetheless fits into one capital-dominated system.

Clearly, the work summarized above does not constitute the entirety of urban studies. Nonetheless, these three areas of theorization – postcolonial 'ordinary city' urbanism; practice theory 'assemblage urbanism'; and Marxist 'planetary urbanism' – are key frameworks that have been used for understanding the shifting global urban landscape. To explore the everyday and the global in relation to the city, however, we cannot just have theories for the city. We need them for the other two concepts, too: that is, for the global and the everyday, or the planet and subjectivity.

Urban machines: environment, society and the self

> Vectors of subjectification do not necessarily pass through the individual, which in reality appears to be something like a 'terminal' for processes that involve human groups, socio-economic ensembles, data-processing machines etc. Therefore, interiority establishes itself at the crossroads of multiple components, each relatively autonomous in relation to the other and, if need be, in open conflict.
>
> (Guattari, 2000, p. 25)

Guattari's argument turns the self inside-out. Arguing against the tendency of psychoanalytical and psychiatric thinking to look at internalized explanations for the self, Guattari looks at those external elements which coalesce in our bodies. This turn to the outside helps Guattari identify three elements that come together to produce human life. He calls these the three ecologies: the environment, social relations and human subjectivity.

I propose that if theories of urbanization are to go global, then they require a conceptualization which connects together earth, social relations and the self. For Guattari, the three ecologies come together produce the assemblage that we could call 'human life or 'global life' (Guattari, 2000). Writing in 1989, Guattari recognizes that what is called 'the ecological crisis' is a crisis not just of the environment but also of individual and collective human life. In an ecosophical approach to subjectivity, it behoves us to think not of subject, or even of subject and environment together, but of dispersed vectors or 'components of subjectification'. As such, subjectivity and the globe are interrelated. Subjectivity can be found in the environment, in social institutions, in landscapes and in internal fantasies (Guattari, 2000). However, these things themselves are also products of subjectivity. Drawing from Bateson, Guattari argues that the earth and the subject are co-produced. As Bateson puts it, the unit of experience is 'organism plus environment', to which the mind, or the subject, is immanent (Bateson, 1973, p. 486)

In a world of cities we can explore different components that make up the urban versions of these three ecologies. The ecology of the self in urban life is characterized by the intensity of our interactions with people, objects, bodies and buildings within these relatively small spaces. Urban space produces a corporeality which is psychological, sociological and physical (Grosz, 2008, p. 430). Simultaneously, the city itself is also produced out of the actions of multiple individual selves. De Certeau (1984) characterizes this production of the city as a fluid, collective, unread narrative, generated by the aggregation of individual actions. Subjectivities associated with the city are often presumed to be more global than others, as can be recognized in the conceptualizations of 'cosmopolitanism' and 'encounter' that are often associated with cities. These two concepts have been used to explore the ways in which cities uniquely throw together an increased diversity of people, producing a range of positive and negative outputs (Wilson, 2017). The globalization of urban life would imply a globalization of both encounter and cosmopolitanism. A key question that this raises in relation to global urbanism, then, is how an increased range of encounters might characterize and shape ecologies of the self.

This globalizing of urban encounters also means that there is a socialized element to the self–world relationship, as is emphasized by Guattari through his inclusion of social relations as one of the three ecologies. From a scalar perspective, the 'social relations' ecology of the city extends from our relations with families, neighbours and communities to the global relations of citizenship, trade and communication that characterize contemporary urban life. These social relations are unequally and unfairly experienced. In other words, the encounters and cosmopolitanism of the urban self are not universally available; rather, they are limited on racial, gendered, sexualized and other grounds. In part, this is because of the role of capitalism, which plays *the* key role for Brenner and colleagues, but is just one of several elements for others (see Derickson, 2015, for a discussion). Cities have spread with the growth of capital, so a focus on the social and the 'global' must attempt to capture this role of capital. Cities have a strategic centrality in connecting capital across the globe (Brenner and Theodore, 2002; McCann, 2010), acting as nodes in global webs. This means that bringing 'earth' into global urbanism means considering the physical links and infrastructures that underlie the transnational networks of capital. However, it also means paying attention to the localized socio-economic inequalities in which spaces of global connection become more and more elitist, as dispossessed communities are forced to urban peripheries. In other words, this is a focus on the small-scale, informal and precarious forms of social interaction that come with global capitalism. To continue the analogy, an ecology of social relations must pay attention to the redwoods and to the weeds.

Finally, Guattari describes the natural environment as the third ecology. His writing was as much a response to the global environmental crisis as it was to social or political transformations. As he writes, 'Lacking sufficient consideration of the dimensions of environmental ecology, social ecology and mental ecology ... humanity and even the entire biosphere will be threatened' (Guattari, 1993, p. 148). This third ecology is in some ways the most important, given that it provides the conditions for living on earth and the existence of the other ecologies. As with the previous ecologies, in a world of cities we need to explore the intersecting local and global environmental elements of both planet and city. On the one hand, environmental crises are perhaps the most obviously *global* element of planetary urbanization. Resource extraction, carbon footprints, transport networks and atmospheric emissions provide simple and clear evidence of the impact that urban society can have on apparently rural locations. Social scientists have come to argue that climate crises must be understand as an inherently urban problem (Bulkeley, 2010). However, environmental ecology also operates at the very local scale. Sites become contaminated with pollutants, river systems lose wildlife, and even where ecological systems are not destroyed by urbanism, they are radically altered (Wu, 2014). We can also identify that ecology shapes subjectivities in ways beyond the obvious interactions between pollutants, environments and the self. Laws (2009, p. 1883) suggests that we need to acknowledge the role of the 'unmappable' spaces of everyday life in which subjectivities are remade, spaces that may or may not be those which are traditionally considered as contributing to the production of selves. In other words, the global urban environment might be considered as a large interconnected ecosphere but also through the application of concepts of niches and domains: small-scale and rather particular environments in which the self is reproduced, for better or for worse.

So Guattari offers us a model for the relations among the social, the environmental and the psychological in his *three ecologies*. We can add to this the concept of the subjectivity machine as a way of understanding how these ecologies come together to produce the people who live urban lives. My argument here is that the city is something of a machine, pulling together the three ecologies to produce selves, places and the world. Machines, as described by Deleuze and Guattari (1984), can be characterized as assemblages that produce. In other words, they are relations of things that together make something new, which have consistency and power. Machines are thus distinguished from other concepts by their relative persistence and stability, though they are always susceptible to changing inputs or components. They operate in relation to and with a diverse range of inputs: 'a technical machine, for example, in a factory, interacts with a social machine, a training machine, a research machine, a commercial machine, etc.' (Guattari, 2006, p. 418). In so far as machines produce, the world itself might be said to be 'machinic'; the final, one-word sentence that concludes Deleuze and Guattari's (1987, p. 566) magnum opus, *A Thousand Plateaus*, is 'Mechanosphere.' These machines are not just biological or technical; they are also implicated in social and power relations: 'the machine has to be conceived in relation to a social body, and not in relation to a human biological one' (Guattari, 2009, p. 100). In turning to the machinic at this point, we can understand that the three ecologies come together in cities to make subjectivities, to make the built environment and to make what we might call 'urban life'. Understanding the relationship between these ecologies as machinic is useful in that it informs us of the ways in which environment, self and social relations *combine*.

If, as a reader, you intend to follow me with this conceptualization of the city, out of either agreement or indulgence, then you might reasonably ask why this is a helpful perspective in the contemporary moment. This also helps me explore the movement towards considering the role of the night in the city. As I mentioned earlier, we have arguably moved to a stage of 'post-globalization' (Gardels, 2008; Peck *et al.*, 2010), in which global connections and capitalism have already spread, and where change is centred on the reworking and remaking of global life rather than on increased connectivity. This process has been fuelled by the spread of forms of capitalism and associated governance that can be broadly described as 'neoliberalism'. Under this philosophy, areas previously left free from market control are opened up under the conviction that such marketization generates economic and social freedom (Brenner and Theodore, 2002; Peck and Tickell, 2002; Harvey, 2007). In a *globalized* world, the role of neoliberalism becomes not one of continuing the geographical spread of capitalism, but one of enhancing the depth and extent of its penetration (Peck *et al.*, 2010). Academics in critical social sciences have repeatedly shown how this deepening and opening of capitalism has left the poor and powerless exposed to the harsher forces of capitalist exchange, while the wealthy and powerful retain and perhaps even increase their levels of protection. Neoliberalism has drawn from the subjectivity machine to try to shape desires and encourage people to want to participate in market practices. Anderson (2016, p. 5) identifies a need to recognize 'the ongoing organization of collective affects and the grip and tenacity of neoliberalism'. In other words, neoliberalism is not just a function of economic and political practice, but of the forms of production of self, society and environment that I have characterized here (see Berlant, 2011, for an influential account of how subjectivity is produced in relation to contemporary capitalism). Once global flows and connections have been broadly achieved, the work of neoliberalism is to deepen and enhance these by developing further control over everyday life. Here, it is worth noting Deleuze's (1992) diagnosis of a move from Foucauldian disciplinary power to what he calls 'control power': that is, from power in institutions to power through the control over everyday actions. For Deleuze, 'controls are a modulation, like a self-deforming cast that will continuously change from one moment to another' (p. 4). Contemporary power in cities looks something like this – not operating through strict concentration of power in the centre, but through the dispersal of more fluid forms of control throughout urban life. In combination with the trends defined above, we see that the production and control of subjectivity in the city has increasingly become the key site of power, governance and struggle.

Night and the limits of the city

In this chapter, I have outlined a vision of urban life which puts forward the city as a machine, through which environment, self and society co-produce one another. The literature cited above draws in a range of different actors as this machine's inputs and outputs, with contrasts to be found between the broadly 'humanist' perspectives of postcolonialism, the emphasis on the centrality of economic exchange in Marxism, and the more diverse pulling together of actants in practice theory (Latour, 2005; Harman, 2009). While paying attention to the claims of these areas, I want to avoid the identification of a 'master' actor, whether that be a particular location, set of relations, concepts or processes. In understanding contemporary urban life and the role of the night within it, we need to see affects, discourses, nature, identities, bodies, materials, animals and emotions as all entering the subjectivity machine equally and together. This means that we do not name a dominant system, relation or process: capitalism, nationalism, sovereign power, patriarchy, (post)colonialism or whatever. Such processes are all important, varying in their nature and effectivity depending on location. However, none of them sufficiently describes or characterizes the relations between the different inputs and the actants that they shape. Instead, such descriptions should appear selectively, from moment to moment and place to place, dependent upon the described relations. Furthermore, my approach does not take any actants as preceding the social and ecosophical relations of urban life. Rather, the characters of the urban world are all taken, at least in the form that they relate to each other, as produced. Therefore, to paraphrase Law and Hetherington (2000), here we are primarily interested not in which actants and relations are produced, but in *how* they are produced. The resulting picture is of a series of never-ending, intersecting processes. Affects shape subjectivities, which in turn produce discourses, which shape identities, which produce new objects, which have affects, and so forth.

All of this is not to deny the powerful analytical insights of theories which have identified the important role of these core processes or positions. Social scientists have worked hard to show the ways that race, gender, sexuality and other forms of social categorization remain axes along which subjectivities are judged and governed. Others have explored how the processes of capitalist accumulation have been powerful forces which have driven changes across economic, cultural and political divides. My aim is broadly to accept all of these arguments, but to overcome attempts to posit any of them as fundamental or paramount. In other words, to deny the ontological status of such categories, or at least to deny the ontological status of the socially produced understandings of these categories, is not to deny their force, power and importance. It is vital to stress that, whatever the city, whatever the social situation, whatever the circulating affects, those who find themselves categorized and placed into positions of little power are likely to have 'worse' experiences of urban space in the form of heightened threats, less freedom, less mobility and so forth. However, the social positions which result in inequalities are themselves constructs, emergent from the machinic city and the intersection of the three ecologies. In relation to power, then, I wish to explore how subject categories are formed, reinforced, performed and circulated, and how they act in consort, rather than identify them as explanatory, causal features in urban relations. This understanding of how a city functions runs throughout this book.

If we argue that the city has globalized and therefore that we need to understand cities in relation to the planet, a possible objection arises about the limits to the city. In other words, is the claim about planetary urbanization the same as the claim that the city is now borderless? Brenner (2014) states, quite unequivocally, that it is: he describes his work as 'urban theory without an outside'. In moving towards night, it is interesting to note that many of the writers in the planetary urbanization framework cite the spread of society into the night as evidence for this destruction of borders. Merrifield illustrates Lefebvre's description of an urbanized world by reference to science-fiction writer Isaac Asimov's planet of 'Trantor': 'From outer space, at night-time … Trantor looks like a "giant conglomeration of fire-flies, caught in mid-motion and still forever"' (Merrifield, 2013, p. 910). Similarly, Brenner (2013, p. 87) focuses on the macro-level images of the earth at night, arguing that these 'illustrate [that] the geographies of urbanization have exploded the boundaries of city, metropolis, region, and territory: they have assumed a planetary scale'. In other words, both Brenner and Merrifield offer evidence for an urbanized world in the claim that the spread of artificial lighting at night provides a twofold visualization of this urbanization. The first part of this claim is that the spread of night-time lighting is indicative of the geographical reach of urban activity, marking as it does the spread of infrastructure beyond the immediate built-up area of cities and their centres. The second element of this claim is that night-time illumination clearly suggests some sort of domination or control over the night, which in this argument stands for 'the natural'. In other words, the work here draws a direct parallel between the spread of human activity into the night and the spread of human activity across space – as have theorists of the night and society, as we shall see (Melbin, 1987).

But this use of night seems to come from above: both Merrifield and Brenner draw on pictures of a planet taken from space. It is clear to see why this god's-eye perspective troubles many writers (Oswin, 2016). I share this concern, and this book is in part an attempt to rescue the night as a 'timespace' of boundaries. We can return to the Chicago School to understand why this is valuable: 'A serviceable definition of urbanism should not only denote the essential characteristics which all cities – at least those in our culture – have in common, but should lend itself to the discovery of their variations' (Wirth, 1938, p. 6).

While we might now be more sceptical of the notion of 'essential characteristics' and of cities being of one culture or not, Wirth reminds us of the fundamental value of precision and limitations in developing understandings of any topic, such as 'urban life'. In other words, beyond the epistemological critique of universalism offered by feminism and postcolonialism, I suggest that there is a practical concern that a concept 'without an outside' is fundamentally useless; or, to paraphrase Adey (2006) on mobilities, 'if the city is everything then it is nothing'. Brenner and Schmid's (2015) 'universality' is based upon the argument that urbanism might be more fruitfully understood as a process, rather than a series of particular empirical facts in the world. They identify a variety of practices and forms associated with these processes, which bring urbanism down to a particular *materialization* of industrial capitalism into particular places and its *transformation* into 'concrete, temporarily stabilized configurations of socioeconomic life, socio-environmental organization and regulatory management' (p. 172). While this seems helpful, it relies upon a Marxist legacy of positioning capital relations somehow prior to other forms of urban living.

As outlined with Guattari, however, the turn to planetary urbanism demands that we have to take the globe more seriously at both its smallest and its largest scales. Embodied urban rhythms are inseparable from planetary urban expansion: 'the phenomenological approach to space and the lived body reveals the two to be inseparable … there are then as many spaces as there are modes of somatization and subjectivation' (Guattari, 1993, pp. 139–140). In other words, as has been argued, even if the city is to be defined as processual, we need to consider a much wider form of processes than those captured under analyses of capitalism. In this chapter, I've attempted to put forward an understanding of 'the urban' as a product of the machinic intersections of capitalist, environmental, social and psychological processes, manifested through the production of self and space. Increased to the planetary scale, the city becomes more earthly in multiple ways – both 'big' in the relations of cities across the globe, and 'small' in the effects on social and ecological niches. City life is typically associated with increased density, diversity and intensity of interactions, but it is unlikely that any set of characteristics can capture something that would cover all cities or all experiences and iterations of city life.

Even if we take to the entire planet to be more or less urbanized, in this book I will turn to the night to explore the city's borders in a fresh way, by looking at a timespace where the processes and practices that constitute the city are more weakly present. Contestation surrounding ideas of planetary urbanism has typically centred on whether there is a physical outside, in the form of either 'the rural' or Global South urbanism, which sits outside of the particular urban systems being discussed (Derickson, 2015; Robinson and Roy, 2015). In contrast, I want to delve into the night to explore if we can find temporal boundaries or limits to the city. This will potentially identify ways in which this understanding of the city – as a series of interconnected processes that are reinforcing one another – collapses. Are there moments when the intensity of urban interactions dissolves beyond what we would label the city? Are there moments when calling people who are isolated in their homes or unable to travel 'parts of a city' simply due to their geographical proximity to others represents falling back on old-fashioned definitions of urban life based on the built environment? The night, as I will argue in the next chapter, is an interesting timespace in which to explore this as it sits tantalizingly at the border between the social and the natural: it is a condition of social relations and of our planetary being.

References

Adey, P. (2006) 'If mobility is everything then it is nothing: Towards a relational politics of (im)mobilities', *Mobilities*, 1(1), pp. 75–94.
Anderson, B. (2016) 'Neoliberal affects', *Progress in Human Geography*, 40(6), pp. 734–753.
Ball, M. (1986) 'The built environment and the urban question', *Environment and Planning D: Society and Space*, 4(4), pp. 447–464.
Bateson, G. (1973) *Steps to an Ecology of Mind*. St Albans, Australia: Paladin.
Bennett, J. (2005) 'The agency of assemblages and the North American blackout', *Public Culture*, 17(3), pp. 445–466.
Berlant, L. (2011) *Cruel Optimism*. Durham, NC: Duke University Press.
Brenner, N. (2013) 'Theses on urbanization', *Public Culture*, 25(1), pp. 85–114.
Brenner, N. (2014) 'Introduction: Urban theory without an outside', in Brenner, N. (ed.) *Implosions/Explosions*. Berlin: Jovis, pp. 14–31.
Brenner, N. and Schmid, C. (2015) 'Towards a new epistemology of the urban?', *City*, 19(2–3), pp. 151–182.
Brenner, N. and Theodore, N. (2002) 'Cities and the geographies of "actually existing neoliberalism"', *Antipode*, 34(3), pp. 349–379.
Bulkeley, H. (2010) 'Cities and the governing of climate change', *Annual Review of Environment and Resources*, 35(1), pp. 229–253.
Davis, M. (2007) *Planet of Slums*. New York: Verso.
de Certeau, M. (1984) *The Practice of Everyday Life*. Berkeley: University of California Press.
Deleuze, G. (1992) 'Postscript on the Societies of Control', *October*, 59, pp. 3–7.
Deleuze, G. and Guattari, F. (1984) 'Concrete rules and abstract machines', *SubStance*, 3/4, pp. 7–19.
Deleuze, G. and Guattari, F. (1987) *A Thousand Plateaus: Capitalism and Schizophrenia*. Minneapolis: University of Minnesota Press.

Derickson, K.D. (2015) 'Urban geography I: Locating urban theory in the 'urban age', *Progress in Human Geography*, 39(5), pp. 647–657.
Farias, I. and Bender, T. (2012) *Urban Assemblages: How Actor-Network Theory Changes Urban Studies*. London: Routledge.
Friedmann, J. (1986) 'The world city hypothesis', *Development and Change*, 17(1), pp. 69–83.
Gardels, N. (2008) 'Post-globalization', *New Perspectives Quarterly*, 25(2), pp. 2–5.
Grosz, E. (2008) *Chaos, Territory, Art*. New York: Columbia University Press.
Guattari, F. (1993) 'Space and Corporeity', *Columbia Documents of Architecture and Theory*, 2, pp. 139–149.
Guattari, F. (1995) *Chaosmosis: An Ethico-aesthetic Paradigm*. Translated by Bains, P. and Pefanis, J. Sydney: Powet.
Guattari, F. (2000) *The Three Ecologies*. London and New Brunswick, NJ: Athlone Press.
Guattari, F. (2006) *The Anti-Oedipus Papers*. New York: Semiotext(e).
Guattari, F. (2009) *Chaosophy: Texts and Interviews 1972–1977*. Los Angeles: Semiotext(e).
Gwiazdzinski, L. (2005) *La Nuit, dernière Frontière de la Ville*. La Tour-d'Aigues: Editions de l'Aube.
Hall, S. and Savage, M. (2016) 'Animating the urban vortex: New sociological urgencies', *International Journal of Urban and Regional Research*, 40(1), pp. 82–95.
Harman, G. (2009) *Prince of Networks: Bruno Latour and Metaphysics*. Prahran, Australia: re.press.
Harvey, D. (2007) *A Brief History of Neoliberalism*. New York: Oxford University Press.
Latour, B. (2005) *Reassembling the Social*. Oxford: Oxford University Press.
Law, J. and Hetherington K. (2000), 'Materialities, Globalities, Spatialities', in Bryson, J., Daniels, P., Henry, N. and Pollard, J. (eds) *Knowledge, Space, Economy*. London: Routledge, pp. 34–49.
Laws, J. (2009) 'Reworking therapeutic landscapes: The spatiality of an "alternative" self-help group', *Social Science and Medicine*, 69(12), pp. 1827–1833.
Lefebvre, H. (2003) *The Urban Revolution*. Minneapolis: University of Minnesota Press.
Lefebvre, H. (2004) *Rhythmanalysis: Space, Time and Everyday Life*. Translated by Elden, S. and Moore, G. London: Continuum.
McCann, E. (2010) 'Urban policy mobilities and global circuits of knowledge: Towards a research agenda', *Annals of the Association of American Geographers*, 101(1), pp. 107–130.
McCann, E. and Ward, K. (2011) 'Urban assemblages', in McCann, E. and Ward, K. (eds) *Mobile Urbanism*. Minneapolis: University of Minnesota Press, pp. xiii–xxxv.
McFarlane, C. (2010) 'The comparative city: Knowledge, learning, urbanism', *International Journal of Urban and Regional Research*, 34(4), pp. 725–742.
McFarlane, C. (2011a) 'Assemblage and critical urbanism', *City*, 15(2), pp. 204–224.
McFarlane, C. (2011b) 'The city as assemblage: Dwelling and urban space', *Environment and Planning D: Society and Space*, 29(4), pp. 649–671.
Melbin, M. (1987) *Night as Frontier: Colonizing the World after Dark*. New York: The Free Press.
Merrifield, A. (2013) 'The urban question under planetary urbanization', *International Journal of Urban and Regional Research*, 37(3), pp. 909–922.

Ong, A. (2011) 'Introduction: Worlding cities, or the art of being global', in Roy, A. and Ong, A. (eds) *Worlding Cities: Asian Experiments and the Art of Being Global*. Chichester: Wiley, pp. 1–26.

Oswin, N. (2016) 'Planetary urbanization: A view from outside', *Environment and Planning D: Society and Space*, online early access: http://journals.sagepub.com/doi/abs/10.1177/0263775816675963 (accessed 17 October 2017).

Peck, J., Theodore, N. and Brenner, N. (2010) 'Postneoliberalism and its malcontents', *Antipode*, 41(1), pp. 94–116.

Peck, J. and Tickell, A. (2002) 'Neoliberalizing space', *Antipode*, 34(3), pp. 380–404.

Robinson, J. (2006) *Ordinary Cities*. London: Routledge.

Robinson, J. (2011) 'Cities in a world of cties: The comparative gesture', *International Journal of Urban and Regional Research*, 35(1), pp. 1–23.

Robinson, J. and Roy, A. (2015) 'Debate on global urbanisms and the nature of urban theory', *International Journal of Urban and Regional Research*, 40(1), pp. 181–186.

Roy, A. (2011) 'Conclusion: Postcolonial urbanism: Speed, hysteria, mass dreams', in Roy, A. and Ong, A. (eds) *Worlding Cities*. Chichester: Wiley-Blackwell, pp. 307–335.

Schivelbusch, W. (1988) *Disenchanted Night*. Oxford: Berg.

Sheppard, E., Leitner, H. and Maringanti, A. (2013) 'Provincializing global urbanism: A manifesto', *Urban Geography*, 34(7), pp. 893–900.

Silver, J. (2014) 'Incremental infrastructures: Material improvisation and social collaboration across post-colonial Accra', *Urban Geography*, 35(6), pp. 788–804.

Smith, N. (1982) 'Gentrification and uneven development', *Economic Geography*, 58(2), pp. 139–155.

Smith, R.G. and Doel, M.A. (2011) 'Questioning the theoretical basis of current global-city research: Structures, networks and actor-networks', *International Journal of Urban and Regional Research*, 35(1), pp. 24–39.

Wilson, H.F. (2017) 'On geography and encounter: Bodies, borders, and difference', *Progress in Human Geography*, 41(4), pp. 451–471.

Wirth, L. (1938) 'Urbanism as a way of life', *American Journal of Sociology*, 44(1), pp. 1–24.

Wu, J. (2014) 'Urban ecology and sustainability: The state-of-the-science and future directions', *Landscape and Urban Planning*, 125(1), pp. 209–221.

2 Fragmenting frontier
Night, time and the city

Light enters the human eye through the cornea. The amount is modulated by the pupil, which is also part of the cornea. The light that gets in is refracted by the lens into an image projected onto the retina, at the back of the eye. These images are then sent to the brain via the optic nerve, and processed into a three-dimensional image of the world around us. This process has two major implications for human vision. The first is that we can only slowly cope with movements between varying levels of light and dark, as the pupil's response to changing light levels is not instantaneous. This produces the 'blinding' effect that we are familiar with from bright light: if part of our field of vision is brightly lit, our pupils respond to limit the amount of light getting in, meaning our ability to see the darker area is reduced. The second is that the 'exposure time' of any image on our retinas is limited to the length of time that we are able to look directly at a view. Outside of exceptional circumstances (see Guenther, 2013), this is a relatively short period of time, effectively correlating with the period of time that we are able to hold our head static. These limitations combine to give humans pretty useless night vision. And this is no surprise as our species evolved as daytime creatures, with our bodies and vision adapted for grassland and scrubland, in which the ability to see over long distances during the day was key to early humanoid survival. At night, our ancestors would retreat to coppices or the edges of forests and sleep in tree branches and later in shelters or caves, possibly with a rotating watch. It is undoubtedly the case that for early humans and the apes from which they evolved, danger and fear were more prevalent at night than during the day. To the present day there are very few societies that have developed an extensive nocturnal social life that could rival that of the day. However, a characterization of night as empty and lifeless is also wrong. 'Nyctophobia' – fear of the dark – 'is not culturally or historically universal' (Edensor, 2017, p. 170). While humans are inherently more limited in darkness, there has always been some ways in which night has been used in society, and there is plenty of evidence of human activity of one form or another at night.

In this chapter, then, we start with the idea of night as a limited but not empty timespace. The history of the night and society tells the story of the various ways in which humans have started to challenge or overcome the hurdles that night presents. Social scientists have told this history as one of 'expansion'. Melbin (1987) uses the metaphor of frontier extensively, drawing connections between the gradual expansion of society over time and space, but he is far from the only one to do so. Indeed, influential writers, including Karl Marx, have identified that the processes of capital expansion have operated in time and space simultaneously (Castree, 2009). I will follow these arguments, but will also point out the moments when the night shows resistance. I will highlight the evidence for those moments in earlier eras when nyctophobia was not so present; furthermore, in looking at the expansion of day, I'll explore those spaces and moments in which the darkness of night persists.

Night in pre-industrial societies

> Night is often only mentioned in relation to ritual operations, attracting the observer's attention through their contents without considering the rituals in the context of the 24-hour day–night cycle.
>
> (Galinier *et al.*, 2010, p. 820)

In studies of 'pre-modern' life, the night has been largely overlooked, too often treated as either a backdrop or a time without interest. Galinier *et al.* (2010) seek overcome this by developing a series of anthropological themes that might be worth exploring in relation to night. For them, as for many who have explored the night, sleep is a key starting point. Most obviously, sleep in modernity becomes increasingly independent of natural light–dark cycles. In pre-industrial societies, there are many examples of sleep changing with the seasons. Bordin (2002) documents, for example, the use of long summer nights for hunting by several Inuit groups, in response to changing diurnal cycles. Ekrich (2006), in his study of the history of night in medieval Europe, shows how sleep patterns slowly evolved into modernity, as clock time became of increased importance over 'natural' times. Interestingly, the 'reason' for sleep is quite unclear, with researchers now broadly in agreement that it seems to serve multiple purposes rather than any single function, and feeling that it may be as tied to temperature as it is to light and dark (Sejnowski and Destexhe, 2000; Yetish *et al.*, 2015). Sleep has several intersecting biological purposes: muscles and joints are repaired; toxins are processed out of the body; growth occurs in children and adolescents. In an intriguing parallel to the nocturnal city, the body at night undergoes the maintenance that is required to proceed through the day. Sleep also serves key psychological functions. Most obviously, it provides the brain with a break, a barrier against new stimuli that allows it to process the activities of the day: in REM sleep, long-term memories are generated out of the short-term memories that have been recently formed. Finally, sleep may confer evolutionary advantages as it keeps animals inactive, hidden and warm while they are in a vulnerable, tired and recovering state. If early humans were able to find a safe place in which to sleep, they could avoid roaming at night while disadvantaged against predators with better night vision.

Despite this, many practices associated with the night can be identified in pre-industrial, pre-capitalist societies globally. The night is a time of myth and ritual. Myths about the creation of the light–dark cycle, about the production of stars and other sources of night-time light exist in most cultures. These myths may pass on knowledge about how to behave at night; they may incorporate descriptions of stars for navigation; they may be used to allow debate and discussion in settings where speakers' identities are hidden, allowing transgressive words or acts, as in the traditional 'night word' songs of many cultures (Galinier et al., 2010). Galinier et al. (2010) argue that night is representative of frontiers and change in ways which make it attractive for ritual – a tradition that continues into the modern world, with various midwinter festivals persisting in the darkness of night or simply the transgression that is permitted on the weekends in the night-time economies of many countries ('They say it changes when the sun goes down', as the Arctic Monkeys sing in one song). Outside of ritual, societies have often taken advantage of night in order to hunt, following the behaviour of animals which may be elusive during the day. Night-fishing is found in a range of locations, from the Polynesian islands (Kirch and Dye, 1979; Hamilton et al., 2012) through North America (Fowler and Bath, 1981) to the Lake Baikal region of Russia (Simonova, 2014). Such practices are emblematic of the fact that while night has always been a time of rest and danger, it has never been a timespace with no activity whatsoever. Rather, night in pre-modern societies has, like day, been a timespace that contains diverse meanings, produces social rhythms in relation to the earth and has seen a globally diverse range of responses. While quieter than day, night and associated darkness cannot be characterized as wholly negative or wholly positive (Edensor, 2017).

In pre-industrial societies, the relationship between night and darkness is firmly established. The night of early modernity did not see significant changes from the night of pre-industrial societies, with darkness still making the night 'a forbidding place plagued by pestilential vapours, diabolical spirits, natural calamity and human depravity' (Ekrich, 2006, p. 56). In other words, Ekrich describes a night full of terrors – from stories of myth ('the topography of nearly every British hamlet was freighted with supernatural importance'; p. 16) through to the more earthly dangers of crime and assault. Authorities sought to control the night through strictly enforced and harshly punished curfews – towns and cities shut down in advance of darkness. Like the societies before them, this temporality was seasonal, such that in the relatively northern latitudes of much of Western Europe, the curfew increased and decreased in length with the days. In rural areas, people had more freedom at night, and the demands of agriculture would mean that work continued, though here people were at increased risk from criminals. Beaumont's (2015) history of those who would ignore the curfew – the 'night-walkers' – explores the range of negative connotations that was put on those who ventured out at night. Restrictions were harsher on marginal groups, particularly Jews and women. However, 'for all of night's dangers, surprising numbers of men and women, either by necessity or by choice, forsook the safety of the family hearth' (Ekrich, 2006, p. 118). Once again, Ekrich reveals a world in which night-time activity is widespread. There was an economy of night-time activity, in which guards or lantern carriers would accompany the wealthy when they went out. 'Night-soil men' cleared the streets of deposited waste or emptied the latrines. Early astronomers worked at night, and the location of the first observatories on urban limits suggests that light pollution was already potentially a factor, as fires or simply reflections from human structures slightly obscured views. The presence of the legislation against night-time activity that Beaumont and Ekrich explore is itself evidence for the desire of various groups to be active at night.

As such, the pre-industrial, pre-capitalist night was not simply a universally empty time. Nonetheless, it was distinct from day, marked by reduced activity, and in places where governmental structures had formed, it was a timespace that was tightly regulated through harsh punishments for any form of activity. From the eighteenth to the mid-nineteenth century, a series of changes occurred which resulted in the transformation of the night into the form that we would recognize in many cities today. Two of these were particularly important: the development of gas and electric artificial lighting; and the emergence of a mechanized capitalist production system.

Electrification

> The authorities' claim to control over the *whole city* through the control of its main arteries is linked to the centrally organized street lighting ... the street lamps stand as signs of this comprehensive claim to power.
> (Schlör, 1998, p. 58; original emphasis)

Schlör's argument is insightful. Street lighting's introduction marked not just a moment of technological innovation or sociological advancement, but fitted alongside other socio-political changes in the eighteenth and nineteenth centuries to give birth to the modern understanding of the operation of the state. Indeed, while less viscerally engaging than some of the examples in *Discipline and Punishment*, Foucault's (1979) argument about the coevolution of technologies of surveillance and control with modern forms of state power could easily be applied to street lighting. This new technology helped facilitate a move from legislation which prohibited night-time activity with rather strong punishments to regulation and control of new forms of nocturnal activity. But if state control were a major effect and aim of increased street lighting, it was certainly not its only cause or outcome, and the history of the contemporary urban night starts properly with these innovations.

Artificial lighting was largely unchanged as a technology until the eighteenth century. The candle was, as Schivelbusch (1988) argues, a remarkably important and long-lasting invention. It separated the site of combustion – the wick – from the fuel – the wax – in a way that fires and torches had not. It allowed flames to be controlled much more carefully, allowed multiple flames to be arranged next to each other in elaborate candlesticks, and burned with relatively low levels of heat and odour. Candles were thus the main source of public artificial lighting for thousands of years, sometimes – as in the great displays in early eighteenth-century Paris – to a massive extent. In the decades prior to the emergence of the gas lamp, technologies had been refined to create oil lamps in which the fuel supply could be altered through dials or switches, showing the need and desire for new and more flexible ways of lighting. The gas lamp, however, was a revelation. As one 1807 magazine had it, gas lighting was 'clear, bright and colourless'; eight years later, a newspaper report stated that it 'completely penetrates the whole atmosphere … [and] appears as natural and pure as daylight' (quoted in Schivelbusch, 1988, p. 14). In a few years, the gas lighting revolution had spread across all of the world's industrialized societies. Descriptions of lighting from the time reveal spectators enthused with glee and delight at this development. One case from Berlin, quoted by Schlör (1998, p. 59), is revealing:

> Yesterday evening we saw for the first time the prettiest street of the capital, which is also our most pleasurable walk, the Linden, in the bright glow of gas lighting. A great crowd of curious onlookers were attracted by this spectacle, and all of them were surprised: for we have never seen the Linden more brightly lit by the most splendid illuminations. Not in meagre little flames but as broad as a hand the dazzling light shoots forth, so pure that one can perfectly well read a letter at a distance of 20–25 paces … Soon the other main streets too will be illuminated in the same way, and Berlin, which is famous for the pleasurable impression it makes during the day, will also agreeably surprise its visitors at night.

The simultaneous socio-cultural developments were extensive. Most obviously, being out in public at night – which had previously been seen as dangerous and/or nefarious – quickly became commonplace. Schlör (1998) imagines the new fascination of someone walking around nineteenth-century Paris, exploring the emerging nightlife that the city streets provided. The image is intriguing, particularly as the night-walker was with these new technologies shifting from the figure of the destitute tramp to that of the whimsical *flâneur* (Beaumont, 2015). Night-time streets became a timespace for strolling and conspicuous consumption in which cafés, bars, operas and theatres opened to unprecedented late hours. We thus see the emergence of a new night-time society. This was not restricted to the capital cities of European countries, either. Lighting quickly spread to the sites of high colonialism and to other key urban centres across the world. The first public gas lights appeared in Bombay (now Mumbai) in 1865 (Woods, 2015); in São Paolo in 1872 (Falbel and Neumann, 2015); in Johannesburg in 1892 (Lee, 2015); and so forth.

Electric lighting followed in the latter half of the nineteenth century and into the twentieth. The *Newcastle Courant* describes Joseph Swan's lecture introducing the incandescent lightbulb rather mater-of-factly: 'Numerous experiments were made to illustrate the speaker's remarks, and were completely successful. Brilliant lights were obtained from the dynamo electrical machine, which was supplied by messers Mawson and Swan' ('Science and Innovation', 1879). Swan encouraged innovation in and around Newcastle, a major industrial centre, and with his help the first electrically powered domestic lights were installed in Cragside, the home of a local industrialist, in 1870. By 1901, Charles Merz had created the world's first electrical grid in the city, connecting multiple homes in a model that was quickly exported so that by the 1920s and 1930s, electric lighting was standard throughout the industrialized world. On a global scale, this electrification was not a simple process; rather, it involved a series of political, social and economic negotiations in which different voices put forward their views as to how electrification could and should proceed (Nye, 1990). Technologies started to move at different rates, with an uneven spread: Baldwin (2012, p. 159) describes the nocturnal streetscapes of US cities in the early twentieth century as 'a crazy quilt of different forms of illumination: arc light towers, arc light gloves near street level, incandescent lights, gaslights with and without mantles, lamps burning gasoline or kerosene'. As the range of lighting technologies increased, this diversity of lighting sources continued to expand. Meanwhile, the initially wondrous electricity quickly became mundane – a feature of the urban background noticed only in its absence (Nye, 2010), a service that residents of cities take for granted.

Despite this spread of light, and the quick transformation of night in the industrialized world, the global process of electrification remains unfinished. In India, for example, electrification has moved in fits and starts through the twentieth and twenty-first centuries (Kumar, 2015b), such that, by 2011, over 403 million of the country's citizens – around 34 per cent of the population – still had no access to any electricity (Palit and Chaurey, 2011). Those who do have access may be limited to just one or two lights per household, with lighting competing with other requirements – most obviously telecommunications – on fragile grids (Kumar, 2015a). Nonetheless, these new technologies of artificial lighting – first gas, then electric – helped overcome the great barrier of vision at night for humans. With the right resources, it is now possible, if desired, to provide daytime-style lighting conditions at night, though later chapters will explore whether such a high degree of artificial lighting *is* desirable. However, the role of technological innovation is only half the story.

Production at night

The historian E.P. Thompson (1967) has argued that what he calls 'time discipline' was imposed and then internalized in the late eighteenth and nineteenth centuries. This marked a shift away from 'task-oriented' temporalities and towards clock-oriented temporalities. Under task orientation, the measurement of time, such as it was, was related to specific activities and seasons, within the capacities of people. Pre-industrial temporality has been characterized as experiential and phenomenological, imposed not by abstract systems but by the temporality of other agents in the world: moulds or bacteria that would destroy crops and food; weather that would shift tasks to particular moments; trade and other relations across distance; and, of course, the cycles of sun, moon, earth and sleep. However, research has shown that pre-industrial temporality was, in practice, more subtle and complex than this: as Glennie and Thrift's (2009, p. 244) historical research reveals, 'clock times were an integral part of everyday life for a very wide range of people in early modern England'. They record, for example, the levels of precision given by diarists and almanacs in the seventeenth century, finding that nine of their sample of twenty-two gave times with a precision of five minutes or better, while fifteen gave times with a precision of fifteen minutes or better. Presenting this alongside a range of other evidence from public records and written descriptions of daily life, they argue that a less precise but still abstract temporality was commonplace across England as far back as reliable records of everyday life are commonplace, suggesting that pre-industrial societies had a level of abstract timekeeping outside of theological and scientific endeavour.

Still, even if non-capitalist societies had a sense of clock time, its power was weaker than it is today. Temporalities of the clock intersected with the task-oriented or season-oriented temporalities that Thompson identifies. One reason for this was the greater agency that natural actors enjoyed on an everyday basis. As noted above, it would be wrong to say that all temporalities were corporeal: human activity had to respond to the temporalities of crops, animals and the seasons. If we were to follow de Certeau's conceptualizations of power, we might say that no dominant set of temporal practices seized power and claimed their *'propre lieu propre'* – that is, their own and proper place. This was set to change.

> The prolongation of the working day beyond the limits of the natural day, into the night, only acts as a palliative. It quenches only in a slight degree the vampire thirst for the living blood of labour. To appropriate labour during all the 24 hours of the day is, therefore, the inherent tendency of capitalist production.
>
> (Marx, 1887, p. 175)

Marx's comments on the shifting temporalities of labour were written with his usual flair and vigour. The era of mechanization of industry provided a huge and rapid shift in the demands of production. In the UK – and indeed across the globe – it was the cotton industry that was the first to implement mass mechanization. New steam-powered machinery, operated by a handful of people, was able to complete the work of hundreds of individual weavers. In a period of just twenty to thirty years, new factories in Derbyshire, Cheshire, Lancashire, Yorkshire and Staffordshire sprang up, generating huge industrial hubs. These hubs attracted workers, and owners quickly realized that they would save a lot of money by operating twenty-four hours a day, rather than switching off the machinery to fit into human patterns. Marx (1887, p. 194) documents how capitalists first stretched the working day to twelve hours, then pushed it further. He also recognizes how the operation of certain industries into the night quickly spilled over into other sectors, even where the specific demands of those areas did not inherently require night-time activity. We thus see the mechanisms behind the 'advance' of the capitalist processes of exchange into the night: a machinic logic spurs changes in key industries, and this is followed by a variety of related services and activities. By the late nineteenth century, this work at night had spread around the world. In the United States, Pittsburgh was the busiest city for late operation; factories worked 'incessantly' from the American Civil War period onwards. The outcome was time cut from its connection to daylight schedules, 'creating spaces dominated instead by the rhythms of production' (Baldwin, 2012, p. 121). This was enhanced by emergent telecommunications networks that would generate demand for work to operate across time zones and thus into the night.

The combination of electrification and the demands of capital worked together to produce the modern urban night. Electrification created new possibilities, with new spaces of entertainment and relaxation and a newly public nightlife on the streets. Here, desire was an important factor, and the excitement of moving into new – potentially dangerous – timespaces was a key feature. But the spread of electrification also facilitated and was facilitated by the spread of capitalism – the new incessancy that the machinery of the Industrial Revolution could tolerate was facilitated by gas and then electric lighting (Schivelbusch, 1988). One person's leisure became another's work, and a new generation of street cleaners, taxi drivers, guides, entertainers, pub workers and others emerged whose employment was now nocturnal. Newspapers and reporting became a near-perpetual industry (Baldwin, 2012), and of course early police and emergency services emerged to cover the new night-time activities, filling the authority gap. While it is simplistic to characterize the pre-modern night as empty of activity, a time of primitive ritual, state retreat and quiet domestic life, it is fair to say that the contemporary night-time city was born out of the transformations that occurred around this time, all illustrated by the plethora of historical work that has analysed this change (Hughes, 1983; Bouman, 1987; Schivelbusch, 1988; Schlör, 1998; Chikowero, 2007; Baldwin, 2012; Beaumont, 2015).

Night as frontier

What sort of night did this era begat? The post-electrification story does not begin at any one moment, in part because the artificial lighting environment that emerged was highly varied. While artificial lighting has become commonplace in industrialized societies, it is not universally accessible, and in many parts of the world its presence is even less pervasive. With regards to the 'gold standard' lighting technology for public lighting, sodium gas emerged as the key technique after the Second World War, producing the familiar yellow glow that became associated with the night-time city. In turn, white sodium started to replace the yellow version in the 1980s, followed by LED lighting over the last ten years. The twentieth-century city at night therefore saw a continuation of the technological developments and trends that had begun in the nineteenth century. It is perhaps not surprising that theorists towards the end of the latter century therefore characterized the night as having been 'colonized' by day (Melbin, 1978; Schivelbusch, 1988; Gwiazdzinski, 2005). Though other authors used this phrase, it was Melbin who really made it the core of his argument, exploring the colonization in two papers in the late 1970s, then elaborating on this work in a book some ten years later (Melbin, 1977; Melbin, 1978; Melbin, 1987).

Melbin's argument is quite straightforward: 'time is a container, and we are filling it in a new way. We are putting more wakefulness into each twenty-four hours' (Melbin, 1987, p. 1). Here, Melbin's conceptualisation of 'big' concepts such as time, night and rhythm is embedded in realist ontologies that much critical social science would reject. We can see the result of this in some of the more hubristic conclusions that he draws from his analysis: for example, in his claim that human 'mastering' of what he calls two unnatural domains for our biology – the sea and the night – is 'a stunning record of enterprising expansions' (p. 3), with no space left for a discussion of the negative ecological consequences of these 'expansions'. Despite these gaps, Melbin's empirical research is insightful, exploring in detail the effects of the expansion of human society. He is the first to make extensive comparison between the spatial and the nocturnal frontier: in particular, he shifts it from a metaphor to a useful analytical tool. Again, though, his analysis varies in quality. On the one hand, his understanding of the spatial frontier universalizes the experience of the American Midwest. He overlooks fairly basic postcolonial critiques of the concept of the frontier, signing up instead to narratives of heroic expansion led by men while neglecting the variety of frontier tales that we find across the world. On the other hand, he draws out a range of analytical insights. The nocturnal population – like that of the spatial frontier – while more diverse than Melbin acknowledges, is still rather sparse and homogeneous in comparison to that of the day. He points out that both frontiers can act as spaces of escape and opportunity:

> the West held out chances for various kinds of freedoms to be enjoyed ... like the colonization of the West, the conquest of the darkness opens a new zone capable of meeting people's needs for escape and opportunity. It offers privacy and fewer social constraints.
>
> (Melbin, 1987, pp. 35–36)

In other words, both timespaces offer freedom to those who are either economically disadvantaged or socially excluded.

While far from flawless, Melbin's analysis includes three key insights that are vital in the understanding of night. These centre on the concept of 'incessancy' – a term Melbin employs when discussing the third of three historic 'schedules' of nocturnal activity. His first 'schedule' is that of rest and sleep, which he associates – in ways that we know are historically simplistic – with the pre-modern world. The second is where 'affairs reach out from both edges of the daylight boundaries' (Melbin, 1987, p. 84) and night becomes a time for relaxation and leisure, or for recuperation and maintenance of the system. The third – incessancy – differs from the second in that night-time activities are now not solely related to leisure or maintenance, but involve a wider range of practices and actors. Melbin does not say that all activities continue around the clock; rather, an organization or a system will continue to spread activities but will do so more diversely. He cites several outcomes of this. The first is that 'the ongoing project has a timing unlike that of the personnel engaged in it' (p. 84). No individual can possibly be present or in charge of a system for twenty-four hours a day. In an incessant organization, power and oversight therefore moves away from the individual and instead becomes either shared, invested in a position, or displaced to tools of monitoring and surveillance. Through a series of case studies, Melbin explores the complexity of these tools, which are used to manage the issues of 'coverage, continuity and control' (p. 85). Second, this generates a rhythm of power in which the location of power and hierarchical forms change according to daily and weekly rhythms. Typically, at night, power moves 'down' a hierarchy. In my own research with street cleaners, the night manager of a street-cleaning team appreciated the greater autonomy in decision-making that he enjoyed in comparison with his daytime managerial colleagues. In this way, even under conditions of incessancy, night might retain some of the independence and established freedoms away from state control and power that it possessed in quieter eras. Third, incessancy is spreading: 'if incessance develops in the workplace, it will soon invade workers' bodies and households' (p. 101). Melbin recognizes that, as well as in work, the incessancy that shift workers experience will be transmitted into home life through changes to domestic routines, and beyond to changes in services which may seek to accommodate incessancy. Families and communities may change as a result. Indeed, this transference of incessancy into the home means that the colonization of night also impacts on day, as sleep and rest are displaced to the daytime. To stretch the idea, we can see Melbin's discussion of the absorption of incessancy by the body and the family as akin to the ways in which European imperial cities were transformed through their experiences of colonialism (Driver and Gilbert, 1998).

This exploration of incessancy helps reveal the transformative effects that the colonization of the night had during the twentieth century. The spread of lighting technologies and the demands created by capitalism opened up night to this expansion. While Melbin does not acknowledge the link, his arguments clearly resonate with Marx's claims that capital would seek to expand across all hours of the day; both authors identify the same processes of expansion.

Writers since Melbin have continued to use the colonization metaphor for understanding this transformation of the night. Gwiazdzinski's work is perhaps the most recent extended exploration of the night as a frontier. The night, for Gwiazdzinski, consists of the intersection of multiple different rhythms; this contrasts with Melbin's singular view of the timespace. Indeed, Gwiazdzinski (2005, p. 20) likens the day–night rhythm of the city to that of the body. In the same way as we often overlook or forget the important psychological and biological roles of sleep, we often overlook the importance of the night-time to the functioning of the city. Empirically, Gwiazdzinski's core arguments about the night are very similar to those of Melbin: it is an overlooked time into which the outcast and the 'deviant' have travelled and in which key maintenance and restorative activities take place. Gwiazdzinski, however, offers a more sociologically sophisticated exploration of night in relation to other social trends and rhythms: he places night within a wider consideration of temporality and the city. Developing his argument in later work, he makes the case for a more temporally sensitive approach to urban planning, in which the capacities of the night-time city are used to produce a more balanced 'temporal ecology' (Gwiazdzinski, 2009, p. 356). Thus, in his work, there is a more holistic vision of the night as part of the wider rhythms of urban life, perhaps reflecting the continued colonization of night that occurred after Melbin published his theory.

Elsewhere, Gwiazdzinski seeks to be much more of an advocate *for* the night, arguing that 'the night has much to say to the day' (Gwiazdzinski, 2007, p. 20). This pushes him to view the expansion of day into night as destructive: there is a danger that day will overwrite night, that we will lose the positive characteristics of nocturnality. The night thus increasingly becomes a contested timespace, in which those who draw from the night and use it seek to protect it in the face of expansionary pressures (Gwiazdzinski, 2005, p. 16). Again, Gwiazdzinski seems to make a more productive use of the frontier analogy than Melbin did, drawing from it to remind us of the hubristic damage associated with colonial expansion.

Gwiazdzinski also outlines a series of potential futures for the night: 'empowerment', in which the night is recognized as a distinct timespace with a series of activities and a distinct population, which gains traction as a core part of the city; 'banalization', in which the night is completely colonized and the night-time city is akin to the daytime city; 'explosion', in which existing tensions surrounding the use of the night-time city continue to grow; and 'harmonization', in which night retains its distinct characteristics but these are cultivated in solidarity with the city of the day. The possible outcomes, for Gwiazdzinski, are thus wide-ranging, and the task of an 'urbanism of the night' is to make the case for night-time cities which are hospitable and open.

Beyond Gwiazdzinski and Melbin, scholars have traced the different ways in which temporalities and urban rhythms have evolved during the twentieth century and beyond, with a greater or lesser focus on the specificities of the night and the city. There are several books' worth of material on urban temporalities to be written, but I want to focus on two influential writers. The first is Lefebvre, who develops Marx's understanding of the spread of capital across time and space in interesting ways; the second is Crary, whose book *24/7* offers a contemporary account of the ways in which urban temporalities are shifting beyond the frontier.

> Night is part of the media day. It speaks, it emotes, at night as in the day. Without respite! One catches waves: nocturnal voices, voices that are close to us, but also other voices (or images) that come from afar, from the devil, from sunny or cold and misty places. So many voices! Who can hold back the flows, the currents, the tides (or swamps) that break over the world, pieces of information and disinformation, more or less well-founded analyses, publications, messages – cryptic or otherwise? You can go without sleep, or doze off.
>
> (Lefebvre, 2004, p. 46)

Lefebvre attempts to track the spread of capitalism into what he calls 'everyday life' – those areas of daily practice which fall outside of spheres of production. Writing from the mid-1930s to the early 1990s, he saw the urban night transformed from a mainly quiet timespace, filled with only a limited number of activities, to a much busier, colonized entity. This pattern reflected the wider trends that Lefebvre saw as capital spread into moments, spaces, times and practices that were previously not subject to its control. In *Rhythmanalysis*, his use of rhythm as a concept is an attempt to identify both change and continuity within capital and everyday life. Repetition is key for rhythm, but is also always connected to difference: there is 'no identical absolute repetition' (Lefebvre, 2004, p. 6). Difference is always introduced into repetition through the very act of repeating; Lefebvre gives the example of the formula 'A = A', in which the second A differs from the first in that it is second. Difference, then, is produced through acts of repetition and this relationship between difference and repetition is key to solving the conundrum of how change can occur where similarity seems to prevail: rhythm is the name for this mixture of both similarity and difference together. Lefebvre develops his argument by showing how capitalism has a particular set of rhythms associated with it: a rhythm of production, in association with destruction, which seeks to impose a constant repetition across different timespaces (p. 55). Lefebvre, like others before him, associates clock time and wage labour with this rhythm. What he produces is an analysis which identifies a spreading capitalism using tools of temporal control and governance as a way of gradually taking hold of those timespaces – such as the night – which fall outside of its power. He seeks a 'rhythmanalysis' to identify both how this spread happens and how existing rhythms may be protected against the spread of capitalism. Working with Christine Régulier, Lefebvre turns to Mediterranean fishing villages for a worked example of rhythmanalysis (Lefebvre and Régulier, 2004), arguing that connections between natural rhythms of tides, geology and animal movements intersect with social rhythms associated with trade, travel and interactions of people across the sea. For Lefebvre, such 'natural' rhythms might be a source of difference, alterity and combat against the dominant rhythms of capital.

While Lefebvre argues that we can see both capital encroaching into night and opportunities for existing nocturnal activities to resist this encroachment through the study of rhythms, the onset of the 24/7 era marks the death of this battle. This is Crary's (2013) argument. His claim is that 'late capitalism' has fundamentally altered temporalities, with the trends highlighted by Marx and others at the beginning of industrial capitalism reaching a tipping point. If industrial capitalism saw the creation of new abstract temporalities based on the logic of machines rather than human or natural timescales, temporality is further dehumanized in a 24/7 society, leading to a 'non-social model of machinic performance and a suspension of living that does not disclose the human cost required to sustains its effectiveness' (p. 9). Crary's work can thus be understood as a continuation of the themes raised by Lefebvre: a chronicling of the continued incorporation of everyday life into the capitalist system through the insidious expansion of rhythms which reject moments of escape, rest, relaxation and solitude. In particular, for Crary, the new logic of 24/7 has seen the death of night as a time for rest and relaxation, the end of night as a time to recuperate from 'the exhaustion resulting from labour or activity in the world, and the regeneration that regularly occurs within an enclosed and shaded domesticity' (p. 22). In Crary's description of 24/7, this loss of rest is a loss of self in the face of incessancy. The move to 24/7 brings to a close what we might call the 'day–night' settlement of Fordist capitalism, which mirrored the welfare and home–work settlements in which a series of timespaces protected workers from the unfettered forces of expansionary capital.

Crary offers not quite the end point in this history of the night, but a good evocation of the dystopian understandings of where we now stand as a result of changes to night, time and society. He describes a situation in which the social rhythms of day have entered the night, disrupting the ways in which night had previously been used as predominantly a time for rest, relaxation and recuperation. Nonetheless, 24/7 does not completely fill the day, as Crary admits. Night remains something different, and perhaps crucially a timespace in which the difference differs for people in contrasting situations.

So while we have a narrative of expansion and domination by capital, it is just that – a narrative. Empirically and conceptually, there are many other stories to tell of the night. In order to move towards the arguments that this book will present with regards to the night-time city, I want to introduce a few examples of the persistence of distinctly nocturnal cultures.

People who go bump in the night: the nocturnals

Workers

As the histories of night outlined above tell us, night has never been a time without work. With the emergence of industrial capital, however, the new machine-powered factories created the first mass employment at night. As well as outlining his theories of capitalism, in *Capital* Marx provides numerous detailed examples of factory operators' abuse of working-day legislation in the nineteenth century, revealing the extent to which night work had spread (Marx, 1887). Through the twentieth century, working at night – typically referred to as 'shift work' – remained predominantly the domain of the working class (Melbin, 1987). In both the era dominated by industrial capital and through the last fifty years, this night work has been male dominated, although there are several industries – such as cleaning and, in parts of the world, manufacturing – in which there is a night-time preponderance of female workers. In addition, domestic work has been increasingly shifted into the night, and a particular trend which has emerged alongside the increased proportion of women working in paid employment is the 'second shift' of domestic work that they do at night (Warren, 2003). As women start to undertake paid work during the daytime, they generally do not see a commensurate decrease in the amount of domestic work that they have to do. Warren describes shift work as one of the key causes of 'time poverty', when households have insufficient time to spend together (p. 737).

The specific work carried out at night has varied. In the UK, research in 2011 by the Young Foundation found healthcare to be the biggest night-time employment sector, followed by maintenance, then leisure and hospitality, with transport, distribution, finance, non-health public services and construction all having significant night-time components. In total in the UK there were around 3.5 million night workers in that year (Norman, 2011). In India and much of South-East Asia, night-time working in technological and Western-oriented service industries is common (Patel, 2006), with people working in accordance with distant time zones. Many globally focused call centres operate three shifts to account for East Asian, European and American markets. In other contexts, night work has a more localized basis, such as in the night markets of Taipei (Su-Hsin *et al.*, 2008), or in agriculture and fisheries, where it may be necessary to respond to the rhythms of animal life. Night workers are more likely to be from lower-income groups, with the range of negative health outcomes associated with shift work endured for slightly higher wages (Weston, 2013). Nonetheless, the demand for flexible – and thus night working – is spreading to the professional, middle classes as well as the working class, particularly in peripheral economic areas and globalized industries (De Angelis, 2010). For example, in the financial sector, global hubs are connected irrespective of local timings.

So night workers are an increasingly diverse range of people, particularly on a global scale, but what links many – though far from all – of them is that they are *pressured* into night working. They are also typically younger, as a group, than day workers. Incessancy means an increasing number of people are working at night, but Melbin's argument still holds true: in general, only a limited section of the population is prepared to undertake this work. Still, studies show that the night is certainly not just the domain of bar workers and criminals. Speaking of which ...

Criminals, deviants and night-walkers

As a timespace of reduced surveillance, night has a long association with crime. Both in research and in popular depictions of criminals, the night is a time in which the protection that the state offers declines or disappears. Worried newspaper reports describe areas of cities at night as 'no-go areas', while late-night opening of bars, restaurants or takeaways is decried as heralding the arrival of hordes of undesirables. Certainly, fear of crime is much higher at night, and researchers have repeatedly shown how this negatively affects women and ethnic minorities – groups who have higher underlying levels of fear of crime (Valentine, 1989; Talbot, 2007). Actual statistics for the timing of crime are difficult to obtain, however. The time of occurrence for many crimes is unclear: reported instances of burglaries in many US cities peak in the evening, but this will also account for crimes that have occurred at any point during the day and are discovered when people arrive home from work. Such difficulties have affected work which has attempted to analyse the role of street lighting on crime (Steinbach et al., 2015). More clear is the occurrence of alcohol-related crimes. In the UK, research has shown that this category of crime – though problematically defined – increases gradually in frequency from around 18.00 to 20.00 before rising dramatically to a peak in the early hours of the morning and dropping off after around 04.00 (Bromley and Nelson, 2002). As such, while evidence is limited and many of the commonsense presumptions about nocturnal criminality may not hold true, a clear socio-cultural association is made between criminals and the night-time city. Criminological research has revealed the diverse range of practices which take place in the night-time city, exploring the geographical and sociological relations that these produce (Measham and Moore, 2009).

Beyond this, it is worth noting that the commonsense association of night with criminality serves to create the presumption that those who are out and about at night are doing something illegal. As Schlör (1998) notes, gathering at night or wandering the streets at night has long been labelled 'dodgy', 'unsavoury' or otherwise indicative of bad intent. Local newspaper reports often contain residents' concerns about developments attracting 'people at night'. As Dunn (2016, p. 77) argues, 'going out into the night can require explanation'. Women are often more likely to be accused of being irresponsible for being out at night, either as sources of trouble themselves or for not showing due care for their own safety (Dworkin, 1993). Historically, marginalized and criminalized groups have congregated in the night. Perhaps the best-researched examples of this are LGBT communities: the night offers the 'cover of darkness' in societies where homosexuality is illegal and where a significant danger of violence precludes social gatherings of LGBT people during the day (Bell, 1991). More recent research has explored the importance of the night for refugees in places where their working opportunities may be restricted by law (Jolliffe, 2016). In other words, the night has been a time into which people have travelled when seeking escape from exclusion and surveillance during the day.

In part because of this, outsiders who have made their way into the night have often been criminalized simply due to their presence. Historically, the act of 'night-walking' itself has been illegal (Beaumont, 2015), although since the nineteenth century it has emerged as a social and artistic practice, associated with a range of literary movements, including psychogeography (Coverley, 2012). A pioneer of night 'expeditions' into the city was Charles Dickens, who documented his habit of walking through London at night: 'The wild moon and clouds were as restless as an evil conscience in a tumbled bed, and the very shadow of the immensity of London seemed to lie oppressively upon the river' (Dickens, 1861). Dickens's night-time inhabitants are lonely wanderers, avoiding the criminal element of the night and allowing the quiet of the city to drive contemplation of the lives of those who sleep around him. Though he experiences the dangers and the discomfort of the night-time city, his time differs significantly from that of homeless people, who do not have the safety that is provided by the ability to return home. In the modern era, as discussed in Chapter 5, 'urbex' urban explorers have been associated with exploring the city at night, venturing into spaces that have been abandoned as well as high-profile sites that have been left unsecured. This practice was perhaps most notably highlighted by Garrett's scaling of the Shard while it was under construction (Garrett, 2013). Such practices reveal new ways of engaging with the city, but they have been criticized for being driven by a masculinist way of understanding the city in which resistance is predicated on corporeal capacity to explore the streets at night (Mott and Roberts, 2013). While this accusation has some merit, it should not undermine the useful insights into urban life that these explorations have revealed. Nonetheless, it is worth reiterating that women have suffered disproportionately for walking in public at night (Schlör, 1998; Beaumont, 2015), and that the night remains, to some extent, a time for 'outsiders'.

Drinkers

Chapter 4 will explore the changing night-time economies of cities in more depth, but for now it is worth noting that groups which are typically labelled as 'party-goers' or 'revellers' in the media inevitably comprise a major presence in many cities' nights. Researchers have focused quite extensively on this sub-group, noting the ways in which those seeking leisure in the urban night are a more diverse bunch than is sometimes portrayed. While it is true that in many cities the night is a timespace where youth is more prevalent than during the day, there is plenty of research which focuses on other groups who use the night-time city for leisure (Jayne *et al.*, 2011). Drinkers and other night-time leisure-seekers – either tourists or residents – have become key targets for urban promoters. An active 'night-time economy' creates employment and makes use of various city-centre buildings that might otherwise be empty (Bianchini, 1995). Led by Amsterdam, several European cities have appointed 'night mayors' to promote and encourage the growth of this sector. Outside of Europe, an active nightlife is often seen as a marker of cosmopolitanism, and key to attracting elites. For instance, Florida (2002) identifies nightlife as a key attractor for the 'creative class'. The night-time economy sits in a fine balance between the mainstream and the outsiders discussed above. While now part of many 'official' urban strategies, drinkers are often associated with various forms of 'deviant' behaviour, as previously highlighted: sexual encounters, alcohol and drug consumption, littering, violence and so forth. Globally, the tolerance towards and extent of such activities vary significantly, and in some locations the urban night becomes associated with particular sub-cultural groups: in Singapore, for example, there is a sharp divide between the city's Chinese community, who inhabit a Western-style nightlife, and its Muslim community (Tan, 2012).

Later, I will explore the ways in which the weekend night in particular plays an important role in the weekly rhythms for young people in many countries. Anthropologists, sociologists and others have explored how the night is a timespace in which young people produce themselves and the city around them, in relation to the habits and routines of older generations (Hollands, 1995). The leisure economy of the night emerged quickly with capitalism and artificial lighting (Baldwin, 2012), and it is now central to many urban strategies. Nonetheless, the relationship between the city and its drinkers is often fraught.

Homeless people

In nineteenth-century France, the '*asiles de nuit*' were created to help deal with the new problem of homelessness. As Katz (2015) outlines in his fascinating history of this period, the *asiles de nuit* were formative in producing a new social grouping; ironically, the creation of spaces of safety and temporary refuge for people without homes contributed to the institutionalization of 'homelessness' as a status and 'the homeless' as a group. The first *asiles de nuit* were opened in Marseille in 1872. Homeless people, as a group, emerged in the UK at a similar time; use of the noun 'homeless' in reference to a homeless person is first documented in 1857. Of course, nomadic or semi-nomadic populations had been in existence long before that date, and the reasons for the emergence of urban homelessness as an issue of concern at that precise moment in time are complex. However, what is interesting is that this social group comes into being at the same time as the extension of society into the night more generally. In the modern era, researchers have focused on the night-time city as simultaneously a space of danger and a space of familiarity for homeless people. Such research has emphasized the geographical knowledge that is developed by homeless people in the city at night in relation to the spaces in which they may safely rest (Cloke *et al.*, 2008) or the routes that they use to travel through and around the city (Jackson, 2012). This community also interacts with all of the nocturnal groups identified above – drinkers, shift workers and night-walkers as well as criminals, whom many homeless people will spend much of the night avoiding.

The nocturnals listed here are just some of those who inhabit the city at night. Furthermore, any analysis of social groups must be alert to the intersectionality of social identities: the experience of a homeless man and a homeless woman, or homeless people of minority ethnicities, will differ. However, these groups reveal that though night may have been colonized, there are multiple nocturnal cultures which seem to be independent of or at least distinct from day. We might, in the terminology of Melbin, refer to these groups as our 'frontier settlers'. Like the frontier itself, they are more diverse and varied than might first meet the eye, though they do represent a population that is more homogeneous than that of the day.

In some ways, then, the following chapters will go some way to complicating and expanding on the narratives presented in this section. First, though, I want to conclude this chapter by exploring the ways in which the known, established night is fragmenting and what this might teach us about the boundaries of the city.

Fragmentation and affects of incessancy on the edge of the city

> The imperative toward just-in-time production of goods and personality meets [with] constant appointment; double booking; postponement; and disappointment, demanding return, repair, and compensation from below ... More and more, even when asleep, one is never closed for business. This is the pressured affectsphere of entrepreneurial subjectivity, the form of life forced to be on the make.
>
> (Berlant, 2014, p. 34)

Berlant describes the lived experience of the incessancy that Melbin, Gwiazdzinski and Crary outline. She uses the term 'affectsphere' where others have used 'atmosphere' (Anderson, 2014) to refer to the socially collated mixture of affects that come together in a particular timespace. The 'affects of incessancy' refer to this collected set of sensations, feelings and pressures that have become associated with night over the last twenty to thirty years. Berlant, like Crary, attaches such affects to the subjectivities developed under neoliberalism, shaped by the demands that this economic and political system makes on people to adapt and bend themselves to be available to capitalist exchange (Larner, 2003; Virno, 2004; Lazzarato, 2015). With incessancy, capitalism combines with the technological capacity for constant operation to make the same demands on the body as are made on machines. It is, in some ways, the final spread of factory logic to daily life. Also notable is that these affects of incessancy – stress, fatigue, expectation, isolation and so forth – are felt among the poor and the wealthy alike: from the demands made of precarious workers to absorb the flexibility of contemporary capital to coercing middle-class professionals to become 'always-on' workers, these affects are spreading around the capitalist system. The circulating affects of incessancy, therefore, could be understood as markers of the colonization of the night. As they continue to spread, they start to reveal the presence of the logic of the city of the day during the night.

However, there are still many stories to tell of places that these affects have not yet reached: of the regular electricity cuts at night in Podgorica; of people watching television at home, alone, without connections to others; of street markets in Taipei; of tourists in Dark Skies Parks; of parents bonding with their newborns; of the many ways in which night persists across the world. This book will relate some, though not all, of these stories. Crucially, it will seek to outline the different ways in which some of them complicate the standard narratives of night, while looking at the boundary roles that the night seems to take.

Chapter 3 looks at the infrastructure that is necessary for the urban night. Focusing on artificial lighting, I explore how – given that night is inherently challenging to human biology – infrastructures have been developed to help society persist through the night. Building on the role that darkness and sleep play during the night, I also explore how it becomes associated beyond the body with rhythms of rest, recuperation and maintenance. Taking, for example, the social rhythms of transport, we might identify a 'night-mode' when either no or a minimal number of services operate. While this night-mode may roughly correspond to actual darkness or the quiet hours, 'the night' as a whole is the intersection of multiple rhythms which enter the night-mode at similar times.

In Chapter 4 I tell the story of what has been called the 'night-time economy', although I make the case for rejecting this term. In exploring night-time leisure, I look to move beyond the excellent overviews of alcohol-fuelled nights provided elsewhere (Jayne *et al.*, 2011) and instead focus on the diversity of night-time leisure practices. Following this, in Chapter 5, I discuss the aesthetics of the nocturnal city: how are cities represented and designed at night, and what norms are associated with this? Here, as I move towards looking at the frontier of planetary urbanism, I try to develop a narrative around the ways in which the social night fragments at its edges.

I push this further in Chapter 6, turning away from the public sphere and towards the more neglected spaces of domestic and personal nights. In so doing, I explore the variety of domestic practices that are shunted to the night, and the ways in which the home at night is a condition for the more public nocturnal city that is studied elsewhere (Blunt, 2005). In looking for boundaries of the city, I question what happens when we become isolated in our homes at night, and the extent to which these timespaces are meaningfully urban.

These arguments should lead me to the exploration of the borderlands of the urban world that I declared to be seeking in Chapter 1. In so doing, my exploration of night is one of a timespace that I label a 'fragmenting frontier'.

This phrase is inspired by applying some of the postcolonial explorations of spatial frontiers to the temporal frontier. If we follow Melbin (1987) and others, the frontier metaphor is a very deliberate choice, seeking to draw out similarities between the spatial and temporal frontiers. Postcolonial theories, with their focus on the complexity and multiplicity of interactions, 'allow one to move from the pastoral and idealized and, above all, from the uniform notion of "the frontier"' (Naum, 2010, p. 106). For Naum, frontiers are characterized by the incompatibility of ways of living, 'where negotiations take place, identities are reshaped and personhoods invented' (p. 107). These negotiations and interactions are affective, and as such contain within them mechanisms of both power and uncertainty: affects are highly powerful tools but also independent and unpredictable sensations. The affectivity of incessancy means that the frontier might be better understood as a 'contact zone', rather than straightforwardly emerging out of clashes between groups. In contact zones, 'subjects are constituted in and by their relations to each other ... not in terms of separateness or apartheid, but in terms of copresence, interaction, interlocking understandings and practices' (Pratt, 2008, p. 7). Affects depend upon bodily contact in some way, so this idea speaks to those times and spaces away from the space of 'high colonialism' (Morrissey, 2005), but which nonetheless exhibit features of expansion, conflict and hybridity. This concept of the contact zone seems appropriate when exploring the type of frontier that we see in the urban night, where various ways of living come together in a space that is different from the 'imperial' centre of the day but nonetheless inherently bound up with it.

Returning to the previous chapter, then, my claim is not that theories about the globalization of the urban are wrong, but that they overlook the fact that even a globalized system can still have edge-lands and borders. As such a frontier, night contains the contradictions, copresence and incongruities that we might expect to find in borderlands. On the one hand, it is possible to tell a fairly linear story: from object of fear, through timespace outside of state control, past a frontier of capitalism, to a current or future point of incessancy. On macro social, political and economic scales, this story somewhat works. It somewhat mirrors the stories that we told in Chapter 1 about urbanization, about the expansion of technology and about other forces that have grown with modernity and capitalism. But, as with the city more generally, the urban night has another series of stories and moments. In this chapter we have met some of the nocturnals who might help tell these stories. As the book progresses, I want to build on these ideas to show how the urban subjectivity machine produces multiple ways of nocturnal urban being.

References

Anderson, B. (2014) *Encountering Affect*. Farnham: Ashgate.
Baldwin, P.C. (2012) *In the Watches of the Night*. Chicago: Unviersity of Chicago Press.

Beaumont, M. (2015) *Night Walking*. London: Verso.
Bell, D.J. (1991) 'Insignificant others: Lesbian and gay geographies', *Area*, 23(4), pp. 323–329.
Berlant, L. (2014) 'On persistence', *Social Text*, 32(4121), pp. 33–37.
Bianchini, F. (1995) 'Night cultures, night economies', *Planning Practice and Research*, 10(2), pp. 121–126.
Blunt, A. (2005) 'Cultural geography: Cultural geographies of home', *Progress in Human Geography*, 29(4), pp. 505–515.
Bordin, G. (2002) 'La nuit inuit: Éléments de réflexion', *Études/Inuit/Studies*, 26(1), pp. 45–70.
Bouman, M.J. (1987) 'Luxury and control: The urbanity of street lighting in nineteenth-century cities', *Journal of Urban History*, 14(1), pp. 7–37.
Bromley, R.D.F. and Nelson, A.L. (2002) 'Alcohol-related crime and disorder across urban space and time: Evidence from a British city', *Geoforum*, 33(2), pp. 239–254.
Castree, N. (2009) 'The spatio-temporality of capitalism', *Time and Society*, 18(1), pp. 26–61.
Chikowero, M. (2007) 'Subalternating currents: Electrification and power politics in Bulawayo, colonial Zimbabwe, 1894–1939', *Journal of Southern African Studies*, 33(2), pp. 287–306.
Cloke, P., May, J. and Johnsen, S. (2008) 'Performativity and affect in the homeless city', *Environment and Planning D: Society and Space*, 26(2), pp. 241–263.
Coverley, M. (2012) *Psychogeography*. Harpenden: Oldcastle Books.
Crary, J. (2013) *24/7*. Los Angeles: Verso.
De Angelis, M. (2010) 'The production of commons and the "explosion" of the middle class', *Antipode*, 42(4), pp. 954–977.
Dickens, C. (1861) *The Uncommercial Traveller*. Oxford: Oxford University Press.
Driver, F. and Gilbert, D. (1998) 'Heart of empire? Landscape, space and performance in imperial London', *Environment and Planning D: Society and Space*, 16(1), pp. 11–28.
Dunn, N. (2016) *Dark Matters*. Arlesford: Zero Books.
Dworkin, A. (1993) *Letters from a War Zone*. New York: Lawrence Hill Books.
Edensor, T. (2017) *From Light to Dark*. Minneappolis: University of Minnesota Press.
Ekrich, A.R. (2006) *At Day's Close*. London: Pheonix.
Falbel, A. and Neumann, D. (2015) 'Sao Paolo', in Isenstadt, S., Maile-Petty, M. and Neumann, D. (eds) *Cities of Light: Two Centuries of Ubran Illumination*. London: Routledge, pp. 74–78.
Florida, R.L. (2002) *The Rise of the Creative Class*. New York: Basic Books.
Foucault, M. (1979) *Discipline and Punish: The Birth of the Prison*. Harmondsworth: Penguin.
Fowler, C.S. and Bath, J.E. (1981) 'Pyramid Lake Northern Paiute fishing: The ethnographic record', *Journal of California and Great Basin Anthropology*, 3(2), pp. 176–186.
Galinier, J., Becquelin, A.M., Bordin, G., Fontaine, L., Fourmaux, F., Ponce, J.R., Salzarulo, P., Simonnot, P., Therrien, M. and Zilli, I. (2010) 'Anthropology of the night: Cross-disciplinary investigations', *Current Anthropology*, 51(6), pp. 819–847.
Garrett, B.L. (2013) *Explore Everything: Place-hacking the City*. London: Verso.
Glennie, P. and Thrift, N. (2009) *Shaping the Day*. Oxford: Oxford University Press.
Guenther, L. (2013) *Solitary Confinement: Social Death and Its Afterlives*. Minneapolis: University of Minnesota Press.

Gwiazdzinski, L. (2005) *La Nuit, dernière Frontière de la Ville*. La Tour-d'Aigues: Editions de l'Aube.
Gwiazdzinski, L. (2007) *Nuits d'Europe: Pour des Villes Accessibles et Hospitalières*. Belfort-Montbéliard: UTBM.
Gwiazdzinski, L. (2009) 'Chronotopies-l'événementiel et l'éphémère dans la ville des 24 heures', *Bulletin de l'Association de Géographes Français*, 86(3), pp. 345–357.
Hamilton, R.J., Giningele, M., Aswani, S. and Ecochard, J.L. (2012) 'Fishing in the dark-local knowledge: Night spearfishing and spawning aggregations in the western Solomon Islands', *Biological Conservation*, 145(1), pp. 246–257.
Hollands, R. (1995) *Friday Night, Saturday Night*: http://research.ncl.ac.uk/youthnightlife/HOLLANDS.PDF (accessed 18 October 2011).
Hughes, T.P. (1983) *Networks of Power: Electrification in Western Society, 1880–1930*. Baltimore: Johns Hopkins University Press.
Jackson, E. (2012) 'Fixed in mobility: Young homeless people and the city', *International Journal of Urban and Regional Research*, 36(4), pp. 725–741.
Jayne, M., Valentine, G. and Holloway, S.L. (2011) *Alcohol, Drinking, Drunkenness: (Dis)orderly Spaces*. Aldershot: Ashgate.
Jolliffe, P. (2016) 'Night-time and refugees: Evidence from the Thai–Myanmar border', *Journal of Refugee Studies*, 29(1), pp. 1–18.
Katz, L. (2015) *L'Avènement du Sans-Abri*. Paris: Libertalia.
Kirch, P.V. and Dye, T.S. (1979) 'Ethno-archaeology and the development of Polynesian fishing strategies', *Journal of the Polynesian Society*, 88(1), pp. 53–76.
Kumar, A. (2015a) 'Cultures of lights', *Geoforum*, 65, pp. 59–68.
Kumar, A. (2015b) 'Energy access in an era of low carbon transitions: Politicising energy for development projects in India'. Unpublished Ph.D. thesis, Durham University.
Larner, W. (2003) 'Neoliberalism?', *Environment and Planning D*, 21(3), pp. 509–512.
Lazzarato, M. (2015) 'Neoliberalism, the financial crisis and the end of the liberal state', *Theory, Culture and Society*, 32(7–8), pp. 67–83.
Lee, C. (2015) 'Ahead and alight in Johannesburg, 1886–1936', in Isenstadt, S., Maile-Petty, M. and Neumann, D. (eds) *Cities of Light: Two Centuries of Ubran Illumination*. London: Routledge, pp. 96–100.
Lefebvre, H. (2004) *Rhythmanalysis: Space, Time and Everyday Life*. Translated by Elden, S. and Moore, G. London: Continuum.
Lefebvre, H. and Régulier, C. (2004) 'Attempt at the rhythmanalysis of Mediterranean cities', in Lefebvre, H., *Rhythmanalysis: Space, Time and Everyday Life*. London: Continuum, pp. 85–100.
Marx, K. (1887) *Capital: A Critique of Political Economy*: www.marxists.org/archive/marx/works/download/pdf/Capital-Volume-I.pdf (accessed 18 October 2017).
Measham, F. and Moore, K. (2009) 'Repertoires of distinction exploring patterns of weekend polydrug use within local leisure scenes across the English night time economy', *Criminology and Criminal Justice*, 9(4), pp. 437–464.
Melbin, M. (1977) 'The colonization of time', in Carlstein, T., Parkes, D. and Thrift, N. (eds) *Timing Space and Spacing Time*. London: Edward Arnold, pp. 100–113.
Melbin, M. (1978) 'Night as frontier', *American Sociological Review*, 43(1), pp. 3–22.
Melbin, M. (1987) *Night as Frontier: Colonizing the World after Dark*. New York: The Free Press.

Morrissey, J. (2005) 'Cultural geographies of the contact zone: Gaels, Galls and overlapping territories in late medieval Ireland', *Social and Cultural Geography*, 6(4), pp. 551–566.

Mott, C. and Roberts, S.M. (2013) 'Not everyone has (the) balls: Urban exploration and the persistence of masculinist geography', *Antipode*, 46(1), pp. 229–245.

Naum, M. (2010) 'Re-emerging frontiers: Postcolonial theory and historical archaeology of the borderlands', *Journal of Archaeological Method and Theory*, 17(2), pp. 101–131.

Norman, W. (2011) *Rough Nights: The Growing Dangers of Working at Night*: https://youngfoundation.org/wp-content/uploads/2012/10/Rough_Nights.pdf (accessed 8 September 2017).

Nye, D.E. (1990) *Electrifying America: Social Meanings of a New Technology, 1880–1940*. London: MIT Press.

Nye, D. (2010) *When the Lights Went Out: A History of Blackouts in America*. Cambridge, MA: MIT Press.

Palit, D. and Chaurey, A. (2011) 'Off-grid rural electrification experiences from South Asia: Status and best practices', *Energy for Sustainable Development*, 15(3), pp. 266–276.

Patel, R. (2006) 'Working the night shift: Gender and the global economy', *ACME: An International e-Journal for Critical Geographies*, 5(1), pp. 9–27.

Pratt, M.L. (2008) *Imperial Eyes: Travel Writing and Transculturation*. 2nd edn. London: Routledge.

Schivelbusch, W. (1988) *Disenchanted Night*. Oxford: Berg.

Schlör, J. (1998) *Nights in the Big City: Paris, Berlin, London 1840–1930*. London: Reaktion.

'Science and Innovation' (1879) *Newcastle Courant*, 7 February.

Sejnowski, T.J. and Destexhe, A. (2000) 'Why do we sleep?', *Brain Research*, 886(1–2), pp. 208–223.

Simonova, V.V. (2014) 'In alliance with a nocturnal landscape: Memory and water law in the north Baikal, Siberia', *Polar Record*, 50(4), pp. 414–420.

Steinbach, R., Perkins, C., Tompson, L., Johnson, S., Armstrong, B., Green, J., Grundy, C., Wilkinson, P. and Edwards, P. (2015) 'The effect of reduced street lighting on road casualties and crime in England and Wales: Controlled interrupted time series analysis', *Journal of Epidemiology and Community Health*, 69(11), pp. 1118–1124.

Su-Hsin, L., Shu-Chen, C., Jing-Shoung, H. and Chung-Hsien, L. (2008) 'Night market experience and image of temporary residents and foreign visitors', *International Journal of Culture, Tourism and Hospitality Research*, 2(3), pp. 217–233.

Talbot, D. (2007) *Regulating the Night: Race Culture and Exclusion in the Making of the Night-time Economy*. Aldershot: Ashgate.

Tan, Q.H. (2012) 'Flirtatious geographies: Clubs as spaces for the performance of affective heterosexualities', *Gender, Place and Culture*, 20(6), pp. 718–736.

Thompson, E.P. (1967) 'Time, work-discipline and industrial capitalism', *Past and Present*, 38(1), pp. 56–97.

Valentine, G. (1989) 'The geography of women's fear', *Area*, 21(4), pp. 385–390.

Virno, P. (2004) *A Grammar of the Multitude*. Los Angeles: Semiotext(e).

Warren, T. (2003) 'Class and gender-based working time? Time poverty and the division of domestic labour', *Sociology*, 37(4), pp. 733–752.

Weston, L. (2013) 'Shift work', in Craig, R. and Mindell, J. (eds) *Health Survey for England 2013*. Leeds: Health and Social Care Information Centre: http://digital.nhs.uk/catalogue/PUB16076 (accessed 25 October 2017).

Woods, M.N. (2015) 'Mumbai', in Isenstadt, S., Maile-Petty, M. and Neumann, D. (eds) *Cities of Light: Two Centuries of Ubran Illumination*. London: Routledge, pp. 37–44.

Yetish, G., Kaplan, H., Gurven, M., Wood, B., Pontzer, H., Manger, P.R., Wilson, C., McGregor, R. and Siegel, J.M. (2015) 'Natural sleep and its seasonal variations in three pre-industrial societies', *Current Biology*, 25(21), pp. 2862–2868.

3 Nocturnal ecologies and infrastructures

In common myth and understanding, night is often presented as *a priori* – outside of human society. As Ekrich (2006, p. 3) puts it, 'night was man's first necessary evil, our oldest and most haunting terror'. It is, in this tale, from the beginning something that is of nature rather than of society, a non-social phenomenon outside of human knowledge, capacity and creation, something lived against rather than with. Perhaps unsurprisingly, the reality of the naturalness of night is somewhat more complex. Social scientists have repeatedly questioned the concept of 'nature': as Morton (2007, p. 15) argues, nature is 'a mere empty placeholder for a host of other concepts ... a Pandora's box, a word that encapsulates a potentially infinite series of disparate fantasy objects'. For Morton, to attach something to the concept of nature is to ignore and overlook the multiple different ways in which it is experienced, lived and 'social'. I do not want to quite demonize the term 'nature' as Morton does – it has a useful, commonsense value to help us speak of the collection of non-human, non-fabricated actors which participate in global life. However, Morton is right to point to the word's role in popular discourse, and he is one of many to explore the limitations of it as an idea, and the limitations of any essential values or truths that are attached to it. As I outlined with the history of night in the previous chapter, there is plenty of evidence for various forms of nocturnal activity in pre-industrial societies. The supposed 'evil' or 'terror' of the night is thus 'socio-planetary': a mixture of invented myth and the real dangers or threats that have been associated with night throughout human history.

In hunting for the boundaries of the urban world, I have argued that the night exists at the crossroads between the natural and the social. In this chapter, I want to unpack further what we mean by the 'natural' night. However, in order to avoid too much use of this somewhat loaded term, I will more precisely (if somewhat clumsily) call it the biogeoastronomical night. In other words, the natural night emerges from the intersection of biological, geographical and astronomical features, with the biological elements having evolved in response to the geo-astronomical aspects. The effect of the biogeoastronomical night has been to produce a series of differences in human society between day and night. Or, rather, it has been to produce a series of differences which mean that, for human society to operate at night as at day, a significant number of socio-technological infrastructures are required to overcome the biogeoastronomical limitations.

Changes in lighting form the biggest differences between day and night, of course, and have been the primary infrastructural challenge. Tasks which are mundane in daylight become significant challenges for humans in the dark. Pre-electrification, the difficulties and dangers of light were limitations on nocturnal human society, and every night-time activity in societies before electrification took place in low levels of lighting. More than just making light easily and quickly accessible, the spread of electric light had other significant benefits. Light became lighter. Disassociated from the smell, heat and danger brought by fire, fundamentally separated from the material conditions of its production, light became easier to access, a resource that could be switched on or off almost instantly. To achieve this at a socially useful scale has necessitated the spread of infrastructural networks that connect cities to power plants and bring cables into homes. In many countries electrification has been the first time that the state has entered the home, the public sphere has entered the domestic, by attaching us to publicly provided or managed networks. Extended beyond homes, networks of street lighting have also been required to allow the city as a wider unit to function throughout the twenty-four hours of each day. Although street lighting is a starting point for this, other infrastructural developments are also necessary for the development of any night-time society. Heating, transportation and communication all need to be readily available. Emergency and other essential services must continue to operate; there is no night-time city without a police force and some form of healthcare. If we were to list these infrastructures that operate at night, we might be able to identify the key minimum provision that is required for urban life to function.

Such infrastructural transformations have inevitably had consequences. In changing the diurnal rhythms of light and dark, humans have altered the ecologies of the environments of multiple other species. The conditions of possibility for the social world to extend into night have destroyed the conditions of possibility for the night lives of many other creatures. Most notable here are the dangers created by artificial light for species which have relied on darkness for cover or excellent night vision when hunting. Other animals have used light as a signal for times of day or season, so mistaking artificial lighting for natural light can cause significant problems. Beyond these issues, light pollution has altered human–astronomical relationships for the general population, amateur astronomers and scientific researchers alike. If the city is a subjectivity machine, as we argued earlier, then this transformation is a major change in the machine's inputs. Research into artificial lighting has revealed implications for human health and well-being, too. Finally, the infrastructures required for human society to operate through the night have, to date, been carbon intensive, with carbon-dioxide-emitting fossil fuels providing the energy required to produce the necessary night-time light and heat.

Through a focus on infrastructure, we see that the relationship between planet and city is potentially confrontational, as the city expands at the expense of various elements of planetary life. Drawing from the political ecology that has emerged from Marxist thought, theories of planetary urbanization have a strong ecological grounding, emphasizing this confrontation. The cover of the *Implosions/Explosions* anthology shows the Athabasca tar sands in Alberta, explicitly connecting this apparently remote sub-Arctic region to the urban life which requires its exploitation for fuel (Brenner, 2014, p. 26). Building on theories of urban hinterlands, planetary urbanization holds that this impact of the city stretches over much larger areas. These sites are reworked by human activity in order for the city to thrive; this infrastructural work is one the ways in which the planet becomes more urbanized.

As we noted in Chapter 1, Bateson (1973, p. 491) has argued that 'the unit of survival [in evolution] is organism plus environment'. In other words, researchers should not ignore the ecological resonances caused by the activities of any organism or species, particularly humanity. In understanding the conditions of persistence of any species, we cannot separate that species from its environment and the relationships within it. Therefore, to develop the ecosophical understanding of the night-time city outlined in Chapter 1, we need to explore the technologies through which humanity is connected to its environment. Our nocturnal ecology is not just of interest on a planetary scale. Within built-up environments, industrial sites, communities and neighbourhoods, 'light pollution' has negative small-scale social and ecological effects (Gaston *et al.*, 2012). As we will explore, changes to lighting levels significantly alter the behaviour of many animals, and can be as devastating as changes to other key environmental features.

This chapter, then, will develop an understanding of how the night-time city operates as an urban frontier by exploring the minimum levels of infrastructure that seem to be required to keep the urban functioning. In so doing, it will attempt to identify the transformative effects of these infrastructures on non-human actors and ecologies, and as such the ways in which the environment must always be embedded within urban subjectivities. The key aim of this chapter is thus to explore how urban life rubs up against the biogeoastronomical night. To do this, first I will consider in more detail the infrastructures that make night possible, before moving on to explore some of these and their ecological impacts in more detail. Given that lighting the infrastructure is most centrally associated with the urban night, I will use this as the key case study for the chapter. So we will consider the history of lighting-related conflicts in the night-time city, and how lighting is being transformed today. The chapter concludes by exploring some of the global challenges faced in providing artificial lighting in a variety of urban conditions, and how these point towards urban borders.

Artificial lighting as a condition of nocturnal possibility

Foucault's vision of the 'conditions of possibility' is developed in *The Order of Things*, where he focuses on the construction of knowledge (Foucault,

1970, p. 182). This text explores the ways in which the conditions of knowledge production may be conceptual, corporeal and material. In other words, for Foucault, there is a relationship between our knowledge and beliefs about what is possible (the conceptual, or discourse), our embodied actions (the corporeal, or practice) and physical infrastructures (the material) that help (re)produce human social life. Foucault's studies of the history of clinics and prisons illustrate how forms of knowledge and the built environment are produced together to create the conditions of possibility for certain ways of behaving (Foucault, 1963; 1979). What makes this part of Foucault's work of particular interest here is the way in which he shows that practice, discourse and the built environment are deeply connected. The conditions of possibility thus relate to dominant forms of understanding as well as architecture and infrastructure. In relation to infrastructure, Graham and Thrift (2007, p. 3) note that these conditions of possibility are not fixed; rather, 'they are transductions with many conditions of possibility and their own forms of intentionality'. In other words, the conditions of possibility which we consider here are not direct causes of the night-time city, but they are one key feature in the production of the urban night. They are fluid and always dependent upon a series of changing socio-political realities. In exploring the infrastructures of the night as conditions of possibility, I argue that their power to contribute to control over everyday practices – remembering Deleuze's development of Foucault, as outlined in Chapter 1 – is always partial and open to change.

What, then, are the infrastructural conditions of possibility for an extended nocturnal society? Before considering lighting, there are others to note. Pre-modern societies had little or no public activity at night, so life was often cut off by the harsh imposition of power: socialization was banned, curfews enacted and a small 'night watch' would typically enforce these regulations. Thus, one of the conditions of night is the development of an infrastructure of state and society which is sufficiently active and open to exercise forms of power that allow for the extension of society into the dark. So the emergence of the contemporary nocturnal city happened alongside the creation of the state-governed disciplinary society. Second, a '24/7' approach to some infrastructures is necessary: electricity, communications, emergency services and various other forms of 'skeleton society' need to remain in place throughout the night, even if the majority sleep. This requires a series of organizational and institutional practices. Third, in most countries, heating will be a necessary feature for at least part of the year, particularly where people are active in public spaces. Fourth, social norms need to evolve so that, rather than a moment for specific activities, night becomes a time or space in which a wider variety of activities and actors are considered acceptable.

All of these features, however, are themselves dependent upon safe, easy-to-access and reliable artificial lighting. This is not just public or street lighting; offices, factories, homes and businesses will all need lighting to operate at night. In other words, while night-time lighting alone is insufficient to provide for an active urban night, it is a necessary condition on which other essential conditions hang. To date, however, this lighting has always relied on several resources: night-time is an intensive period in the city's 'metabolism', when considered in terms of the amount of energy required in relation to the amount of activity that occurs. That is, lighting uses a lot of energy and produces a lot of waste as part of the 'social and bio-physical dimensions to urban space' (Gandy, 2004, p. 364). While, as we shall discuss, this metabolism need not be inherently carbon-burning, it will always have a high energy-to-activity ratio.

Light pollution, carbon use and the ecological impacts of the night-time city

Light pollution emerged as a problem with the introduction of mercury- and later sodium-based outdoor lighting. These supplanted the earlier incandescent lighting bulbs as they were brighter and more efficient, resulting in rapid increases in both the quantity and the quality of outdoor lighting (Riegel, 1973). The spread of concern about light pollution can act as something of a proxy measure for how this form of lighting spread: I can find no reference to 'light pollution' in any publication before 1970, but it was in widespread use five years later. The first published mention of it that I have found is in a list of potential problems created by poor urban planning (Branch, 1970, p. 724); only a very brief explanation of the term is provided, however, so it is hard to believe that this was the moment when the term was coined. 'Light pollution' makes its debut in the *New York Times* archive in 1971 (Cornell Jr, 1971), yet just two years later the term is sufficiently well known that in the paper cited earlier in this paragraph, Riegel (1973) does not feel the need to introduce it to the generalist readership of *Science* magazine.

Light pollution was first noticed and raised as a problem by astronomers in California, but it quickly spread across the United States and into other nations, alongside the diffusion of new outdoor lighting technologies. Broadly, light pollution can be split into three different kinds. The first, which we might call *astronomical* light pollution, is lighting which prevents the observation of astronomical features. To the naked eye, a very small amount of light can obscure some stars, so throughout history people must have distanced themselves from others in order to get a good view of the night sky. (Consider the siting of pre-industrial observatories on the outskirts of cities.) While telescopes can withstand significantly more light pollution than the human eye, any restrictions to observation may have impacts when studying distant astronomical bodies. This form of light pollution was also the first to be identified, and while light pollution today is a major issue for several reasons, it has long been the astronomical community, both amateur and professional, that has led concerns over the phenomenon (Bogard, 2011). Today, the residents of many Western cities are in effect unable to see the constellations in the night sky, with only rare pockets of darkness revealing the brightest stars. For many people, in the space of a single human lifetime, this has significantly changed the way in which they experience the night.

Elements of *ecological* light pollution – the second form of the problem – predate the coining of the term, but have been recognized as a coherent set of issues only in recent years. Certainly, there has been a much longer history of recognizing that artificial light can have significant effects on the behaviour of living organisms: in 1940, for example, researchers explored the impact of both natural and artificial changes in illumination across multiple disciplines (Park, 1940). Among humans, light therapy for various diseases also has a long history; indeed, Niels Finsen was awarded the 1903 Nobel Prize in Medicine for his research into phototherapy treatments for smallpox. The study of the effects of excess artificial lighting at an environmental or ecological level, however, remained surprisingly limited through much of the twentieth century, with biological research focused on laboratory conditions or the responses of individual species, rather than the real-world impacts of artificial lighting. (Longcore and Rich (2004, p. 191) were the first researchers to coin the phrase 'ecological light pollution' in their summary of existing research, defining it as 'chronic or periodically increased illumination, unexpected changes in illumination, and direct glare' in an environment. A variety of ecological problems emerge from this, but key to them are the disorienting effects which change foraging, reproductive and communication behaviour. These have major influences on inter-species behaviour, particularly hunter–prey relations, which is why ecological light pollution needs to be understood at the systemic, environmental scale, rather than just through the specific biological impacts of artificial lighting on individual organisms (Gaston *et al.*, 2012). Still, a range of particular species-level examples illustrates this wider environmental impact. Crows, monarch butterflies and silver eels are among the species to have had migration patterns disrupted by artificial lighting; young sea turtles, which use the brightness of the sea as a guidance tool after hatching on beaches, are disoriented by bright lights; frogs and mockingbirds have had mating rituals disrupted; and predators such as bats and owls have learned to make use of artificial lighting when hunting insects and rodents, potentially disrupting ecosystems through over-hunting (Navara and Nelson, 2007). These examples resonate around a range of ecosystems to produce spiralling negative and unanticipated effects of artificial lighting.

Relatedly, the third form of light pollution, which we might call *anthropological* light pollution, is also biological in its origins. Here, light pollution can be considered not just in relation to public night-time lighting but also with regards to exposure to excessive artificial lighting in the home. Research into the impact of artificial lighting on human biology has a much longer history (e.g. Knauer, 1980 is a seminal paper) than the study of the broader biological, psychological and social effects of excessive light on humans at a community level. As with ecological light pollution, research in the twenty-first century has explored how light pollution has a series of cascading negative effects. Artificial lighting has now been associated with a variety of serious health problems, including depression, sleep disorders and even certain cancers, such that Pauley (2004) describes it as a 'public health issue'. In a Western context, light pollution is greatest in the inner cities, and may thus contribute to urban populations' sense of alienation from the natural environment, particularly among those who for financial or other mobility reasons are less able to travel to rural areas (Hölker et al., 2009). An emerging component of anthropological light pollution is the proliferation of screens, particularly in the home, where artificial lighting is being encountered for increasing parts of the day.

Therefore, across these three forms of light pollution, the problem is now so pervasive that its effects are felt from the scale of cancerous cells in the human body to that of the visible universe. In other words, the impact of the continuation of night-time society is multi-scalar, revealing the importance of considering these conditions of possibility of night.

The environmental impact of the artificial lighting on which night-time society depends does not, however, stop at light pollution. Most obviously, artificial lighting currently uses a significant amount of energy. One estimate suggested that global public street lighting used 114 TWh (terawatt-hours, or 10^{12} watt-hours) in 2006 (Gaston et al., 2012), which equated to approximately 3 per cent of global electricity production in that year. However, this estimate excluded the privately owned lighting of shopping malls, garages, skyscrapers, bars, offices and factories, as well as the private lighting and energy used in off-grid night-time lighting. For public bodies, lighting is a significant proportion of annual expenditure and energy use, at its highest for those in rural areas in high latitudes.

In many countries, domestic energy use peaks during dark evenings. Research by Powells et al. (2014) shows the daily energy use patterns in homes in northeast England. In this region, energy use is at its highest between 16.00 and 20.00 in December and January, when people are at home in what we might call 'night' conditions: that is, when it is dark, cold and people are not working. We see a near doubling of energy between evenings that are 'day-like' (light and warm) in June compared to those that are 'night-like' in December. Here, then, it is the combined energy demands of lighting and warmth that are important. Indeed, in the difference in energy use between 17.00 in the evening in June and December at these latitudes, we effectively see the difference in energy required for people to be active during the day and for them to be active at night. In other words, we might expect an hour of activity at night to use nearly twice the energy of an equally active hour during the day. There are two caveats to this, however. First, after the evening lighting peak, decreased activity reduces energy use. In other words, the intensity of night-time activity does not counteract the relatively thin spread of it with regards to overall energy use. Second, considered globally, the impact is somewhat varied: for instance, in some countries, such as Australia, the widespread use of air-conditioning means that the day is more energy intensive than the night. Nonetheless, on a global scale, the lighting of night is a major component of public energy use in particular, and cannot be eliminated through small changes in behaviour.

Perhaps unsurprisingly, there has been criticism of artificial lighting, led by the 'International Dark Sky Movement'. This has been driven by the astronomical community (Smith, 2009), and most recently has centred on the development of 'Dark Sky Places'. Such a move has commodified dark skies, with areas now rushing to declare themselves Dark Sky Parks, Dark Sky Communities, Dark Sky Reserves or Dark Sky Sanctuaries in order to attract relatively wealthy dark sky tourists. Their geographical spread is indicative: as of mid-2016, only five of the fifty-seven officially designated Dark Sky Places around the world were outside the United States and Western Europe, and only two were in the Global South (the AURA Observatory in Chile, a privately owned site run and managed by forty American universities; and the NamibRand Nature Reserve in Namibia). While, clearly, it could be said that dark sky protection needs to occur first in those regions where skies are at their most polluted, perhaps the most threatened skies are those above the rapidly urbanizing regions of Asia and Africa, where megacities are growing exponentially. Therefore, while the International Dark Sky Movement should be commended for protecting the ecology and astronomy of certain regions, its capacity to limit the spread of artificial lighting will remain questionable while it continues to focus on sites that are visited by wealthy amateur astronomers and naturalists.

Another urban-focused movement, organized by the World Wide Fund for Nature, is 'Earth Hour'. This involves the coordinated switching off of public and private lighting across urban centres; by 2015, cities in 172 countries had registered with the scheme. Here, the focus is on lighting due to its symbolic value, with the campaign delivering a wider message about the environmental impact of energy use. Earth Hour gets its message across by contrasting uncanny images of darkened city skyscapes with the much more familiar images of brightly lit urban environments. The darkened urban night is used to highlight the extent of the artificial lighting network, and this imagery reveals how reliant urban life has become on this condition of possibility. However, other than raising awareness of the issue, Earth Hour has had little impact on light pollution, and the same could be said of similar movements.

In this section, I have argued that any society which continues running during the hours of darkness requires a significant amount of energy. At present, we supply this primarily through burning carbon-dioxide-emitting fuels. Thus, one of the conditions of possibility for night-time urbanism is the burning of fossil fuels to power the artificial lighting and heating networks. In turn, this means that, as part of the 'urban metabolism', night is a particularly energy-intensive time. At the moment, the night does not have the same levels of public or domestic activity as the day. Any increase in activity, however, will have a high energy impact, which is a cause for local and global environmental concern.

Changes to urban lighting and global challenges

Despite some of their negative impacts, lighting technologies are multifaceted and evolving rapidly. These transformations have significant implications for future global lighting and as such for the impact of the conditions of possibility of the night. As outlined in Chapter 2, the electrification of cities was a major innovation that fundamentally altered the nature of cities. It intersected with growing populations, with increased leisure time and with mechanization to help drive the Industrial Revolution and associated urban growth (Schivelbusch, 1988; Schlör, 1998). Although there has been a continued evolution in technologies, as Nye (2014) points out, this has not been a linear process across the globe, and in many places we can still find streets and homes lit by older technologies. Furthermore, a range of other lighting types, including the neon lights of advertising and metal halide floodlighting, has intersected with the classic yellow sodium glow of the city (Jakle, 2001). City lighting has never had a singular format, and as quickly as it was introduced lighting has been unequally present and accessed. The politics associated with lighting varied according to time and place. In countries, cities or regions where lighting has been well provided, or where access has had minimal restrictions, lighting quickly becomes an apolitical part of the urban background. In other times and places, however, lighting can be a highly political assemblage, with failures in provision quickly attacked as failures in government. Decisions about when, where and what to light can also be highly controversial. Like many infrastructural systems, lighting recedes into the background only to reappear at key moments when it becomes contested.

Early lighting controversies emerged from Leveller-style movements that sought to destroy artificial lighting. In both Paris and London, the first lighting brought protests from 'night-walkers' (Beaumont, 2015) who felt that artificial lighting would remove the freedom enjoyed by people at night outside of the surveillance of the day. From an opposing perspective, moralists at the time were concerned about these leisure behaviours and protested that nocturnal congregation might facilitate all sorts of immoral activities (Schivelbusch, 1988). However, as technologies stabilized and spread, lighting began to be seen as unproblematically good – a force for progress and modernity (see the various chapters in Isenstadt et al., 2014). While Nye (2010) shows the ways in which blackouts have been major political events in the United States and others have explored that the failure to provide electricity and therefore lighting has long been a sign of poor governance in many developing countries (Osayimwese, 2014), the controversy has been over failures to provide electricity or lighting, rather than over the nature of the lighting itself. Indeed, as Kumar (2015) has argued, access to public lighting can be understood as recognition by the state, as a claim to legitimacy made by both the state itself (as provider of service) and citizens (as receivers). In all of these later cases, then, controversy over lighting has centred on its unequal provision or unavailability rather than over the nature or existence of the lighting itself.

Since the mid-2000s, however, the provision of public lighting has become newly controversial. There have been multiple sources for this politicization, but key to them has been the creation of new LED lighting technology, which seems to have the potential to solve a wide range of night-time lighting problems. In particular, LEDs seem to offer solutions for both austerity- or carbon-conscious urban authorities in the Global North, and for cost-conscious and resource-poor authorities in the Global South. They have several attractive characteristics. First, they can be turned on and off more quickly and with much greater efficiency than previous technologies. This increases their flexibility in comparison to vapour-based lighting, which takes time to warm up. Second, and building on this quality, LEDs are better suited to various forms of 'smart' city technology, as they can be manipulated in real time. Emerging smart lighting technologies use a mixture of control-centre-monitored responsive lighting, pre-programmed lighting schedules and responsive lighting based on both data and sensors. Third, LEDs are more energy efficient than other technologies, requiring less energy to produce the same levels of brightness. As well as reducing carbon footprints when embedded in traditional networks, LEDs are therefore better suited to power sources such as solar batteries or small-scale generators, again increasing the appeal of the technology. Fourth, LEDs can produce light in a range of different colours, including a much 'whiter' light than vapour- or fire-based lighting, which renders objects more visible at a lower energy level. This makes LEDs particularly suitable for festivals or high-profile lighting sites as well as everyday lighting.

Cheaper, greener, smarter and more flexible – what's not to love? LEDs apparently offer the ability fundamentally to alter the conditions of possibility of the urban night. While, to date, the urban night has been reliant on high-cost and high-energy lighting, LEDs *promise* to reduce this level of demand. Unsurprisingly, then, local and urban authorities across the globe have begun to explore the installation of LED lighting. Many have used LED lighting to introduce schemes which either dim or switch off street lighting for part or all of the night, with lighting levels responsive to monitored street usage. Such schemes have spread quickly. Research by APSE, a collection of local authorities in the UK, found that the reported costs of street lighting were beginning to fall – a change which the study attributed to the introduction of dimming and switching-off schemes over the previous five years (Bailey, 2017).

More ambitious smart-lighting projects are currently largely confined to very small-scale areas or the ambitious plans of lighting and tech companies. One of the largest and oldest of the implemented schemes can be found in Oslo, which has had smart lighting since 2011. The brightness of the city's street lights responds to the time of day and to weather conditions. Lanterns over pedestrian crossings or junctions can be set to greater brightness than neighbouring lanterns, and the system can be overridden to provide extra lighting during major events or incidents (CISCO, 2014). As the brightness levels of lanterns are monitored, the city authorities can identify the exact amount of lighting that is being used at any given moment. Although Oslo's lighting initially included 'dynamic' lighting – using sensors to respond to traffic and pedestrian levels – these were withdrawn due to failures and errors in the technology. Also, the city's system does not yet allow lighting levels of individual lanterns to be controlled dynamically, so any differences within a group of lanterns have to be pre-set.

Despite this innovative practice, at the time of writing, public use of dynamic lighting seems to be limited to small experimental areas, including Gordon Street and the Clyde Riverside Walk in Glasgow; car-park lighting at Oulu Airport, Finland; and an outdoor lighting lab in Albertslund, Copenhagen. Such schemes have proved that sensor-based, dynamic lighting is possible, but as yet schemes which scale this up to the neighbourhood or city level remain conceptual rather than actualized.

LED lighting has also spread quickly outside of the Global North. It has been introduced particularly in areas experiencing rapid urbanization, where the low costs of the technology are beneficial. Adkins *et al.* (2010) show the savings that are possible in homes that use solar-based LED lighting, with villagers in their research area reducing their lighting expenditure by an average of 85 per cent within a month of installing LEDs, as well as obtaining significantly more hours of light. Rapid development in solar-battery technology promises to enhance these benefits. Researchers have reported on lighting initiatives using solar-based LED lighting in locations as diverse as Tibet (Pode, 2010), South Africa (Lemaire, 2011), Kenya (Alstone *et al.*, 2014) and Yemen (Pode, 2013), with the technology favoured by major development organizations such as the World Bank, whose 'Lighting Africa' project has pushed solar/LED as a key method of rolling out lighting to both urban and rural areas. Further innovations have moved beyond LED. The most notable of these is the intriguing 'Litre of Light' project. The technology behind Litre of Light is simply an old plastic bottle filled with bleach and water or chlorine and water. Placed in the roof or wall of a building, this refracts sunlight with the same lighting level as a 40–50-watt bulb. Litre of Light has obtained coverage and funding through a series of art and performance projects at events across the Global North, appearing at major lighting festivals, and has used this to drive installation of the technology in the Philippines, India, Colombia and elsewhere. Such technology is cheaper than LEDs and provides indoor lighting in built-up areas, but it is obviously limited in its ability to light the *night*, as all lighting is lost after darkness falls.

Despite these apparent benefits of new technologies, a number of challenges remain. Projects to spread artificial lighting in the Global South still fall way short of providing public lighting at Global North levels – revealing gaps in urban systems, even where infrastructure may appear to be rolled out. Kumar's (2015) research into the electrification of villages in Bihar finds the provision of lighting to be exacerbating existing gender and caste divisions. Networked lighting is provided typically to one porch light per household, with the remainder of the property still lit by kerosene gas and no public lighting provided. In Kumar's case study, men would congregate in the lit areas, with women allowed only in the poorly lit rooms at the back of the homes. Undoubtedly, the lighting provision is a long way from being anything other than unequal. In the Global North, the introduction of flexibility and variability in lighting levels may also represent increased inequalities, with little certainty as to the social effects that smart lighting might have. A major research project in the

UK in 2015 found no correlation between changes to street lighting and crime or road traffic accident levels (Steinbach *et al.*, 2015), although this study does not rule out possible impacts in particular locations, or impacts on fear of crime.

Furthermore, the social impact of lighting should not be understood as limited to these two measurements, important though they are. Humanist urbanists in the tradition of Jane Jacobs and Jan Gehl have argued for the benefits of equal access to public space throughout the day. While actual levels of crime and danger may not appear to be affected, and smart lighting promises adequate lighting for social interaction, the capacity to deliver on this is currently unproven. A significant concern is if reduced lighting levels encourage people to believe that streets are more dangerous or less accessible, they may reduce night-time activity regardless of actual changes in danger (Green *et al.*, 2015). This may intersect with the perceived 'harshness' of white lighting, particularly among people who are used to sodium's yellow glow, to reduce night-time mobility among some groups. Put differently, to date there has always been a strong correlation between increased night-time lighting levels and night-time activity, and we should be suspicious of smart lighting's claims to overcome this correlation where lighting is reduced. In the Global South, LED lighting is to be welcomed, but there is little or no evidence that it has had any significant impact on existing developmental inequalities. Finally, the spread of smart lighting continues a transfer of power away from elected or appointed officials and towards automated routines, rhythms and algorithms. In other words, while intentions may be positive, transferring the decisions over lighting provision away from elected or appointed officials and to computer programs and/or the people who have been trained to operate them may be anti-democratic.

LED and smart lighting have the potential to transform both the experience of and the conditions for operating the night-time city; it should be clear that there are many benefits, and that they have the capacity significantly to reduce the ecological and environmental impact of night-time urban activity. As things stand, the spread of LED lighting seems almost inevitable: governments and local authorities are strongly supportive, and technologies are likely to improve and proliferate. Furthermore, as has been noted, the spread of LED is unlikely to happen universally, and public political debate is likely to move away from doubts over its merits and towards demands for obtaining its benefits for all areas of the world. While much of the world struggles to obtain adequate and accessible lighting, there are certainly many benefits to this technology. However, it is for these reasons that nocturnalists need to be concerned about LED and smart lighting. New technologies politicize – they create moments for debate, change and innovation. No technology is inherently transformative, for either good or ill. There is a danger that LED technology will be viewed as a panacea and introduced without question across the globe. This may result in bad, unequal and undemocratic lighting provision. At a time when it is still possible to question and moderate the amount of lighting provided in cities at night, an underscrutinized global roll out of LEDs has the potential to create ghettos of artificial light, accessible only to the few, yet simultaneously undermining attempts to reduce overall brightness levels in cities.

The broader story of artificial lighting exemplifies the expansion of day into the nocturnal frontier. The literature on planetary urbanism uses images of street lighting on a global scale to show both the extent of and the gaps in the urban built environment (Brenner, 2013, p. 87), reflecting the broader discursive power of artificial lighting to represent human development (Pritchard, 2017). Such images of a global network of lighting hide as much as they reveal, however. At the planetary scale, focusing on the light obscures the dark. When we delve into the city, we find a mixture of lighting technologies (Baldwin, 2012), and discover that these technologies are neither equally bright nor equally accessible. In other words, even this key form of expansion is partial. Returning to Melbin's (1987, p. 3) claim that human 'conquest' of the night has been 'a stunning record of enterprising expansions', we can identify that while there has indeed been an impressive expansion, this hubristic description overlooks two crucial factors. First, it fails to acknowledge the gaps and inequalities of lighting, as described above. Second, it overlooks many of the negative impacts of different types of light pollution caused by the rapid spread of artificial lighting across the earth during the twentieth century. In summary, therefore, the presence of artificial light does not necessarily equate to a 'conquered night'.

Beyond lighting: infrastructural frontiers at night

Lighting and energy provision are not the only infrastructural practices required for the night, and it is worth turning briefly towards another set of practices – the role of cleaning and maintenance in defining the night-time city. Night-time city streets are often dirty places that need significant cleaning and maintenance. The cleaners who work on the streets or sanitize offices at night form part of what appears to most people – though, of course, not to the cleaners themselves – as the urban background that keeps society ticking along. In one of my favourite social science quotes, Goffman (1974, p. 1) makes the point that

> all the world is not a stage – certainly the theatre isn't entirely. Whether you organize a theatre or an aircraft factory, you need to find places for cars to park and coats to be checked, and these had better be real places, which, incidentally, had better carry real insurance against theft.

To follow Goffman's phrasing, if you are going to construct a major urban development, then you are going to need cleaners, and you will probably want them to operate through the night if people continue to be active then, or unless you want your daytime city users to be disrupted. Like Goffman's car parks, cloakrooms and insurance, however, such cleaning and maintenance is often overlooked in social scientific analysis of places and organizations. Yet, as Melbin's comments suggest, these activities are not peripheral; they are essential, a key way in which society rejuvenates itself.

In my own research with street cleaners in Newcastle upon Tyne, the cleaners were fully aware of their simultaneous necessity and invisibility. Their manager reported – with some frustration – that from time to time, new councillors or local authority officials would question the expenditure on night-time cleaners. His response was to create visibility in two ways. First, he would invite anyone who showed an interest to join the cleaners for a night. Certainly, from my own ethnographic work, such an experience is not easily forgotten and quickly reveals the necessity of the task. Second, he habitually photographed his cleaning team at work, so he had folder upon folder of photographs documenting the different types of cleaning that they did, and the sites that they visited. His photography, alongside my own observations, revealed a city covered in a range of materials at night: vomit, urine, fast food, fast-food packaging, leaflets, bottles, more leaflets, condoms, grease, clothes. Newcastle, like other towns and cities with alcohol-based night-time economies, is constantly strewn with the debris of this timespace of excess. Here, the detritus left on the streets in front of clubs, beside taxi ranks and trailing along the routes that people follow while walking around city centres can be understood as the physical embodiment of the excess of emotion, affect and alcohol that flows around the night-time city.

Beyond the need to remove this material from the street, the urban night is often used for cleaning and maintenance. As Melbin (1987, p. 83) puts it, 'the new order of the day is to rely on night-time to restore the community's well-being'. In offices, factories and transport hubs, an army of cleaners arrives and clears out the day's waste. Whether it's an evening shift following closure, an 05.00 start before the rest of the workers arrive, or an all-night job in which a night shift of maintenance workers replaces a day shift of office staff, the night acts as a timespace in which many buildings are 'reset' and returned to the status of the day. Cleaning is just one form of urban maintenance that is prevalent at night; as noted previously, construction and maintenance is one of the key night-time industries (Norman, 2011). Other key maintenance activities are often nocturnal: work on transport infrastructure, restocking shops, servicing telecommunication systems and work on various public buildings all routinely take place at night. Some forms of infrastructure are switched off at night in order to be reset and restored for the next day. All of this activity pushes us towards an understanding of the night-time city as presented by Crary – that is, as a 24/7 system in which the logic of constant activity has been absorbed. Even though activity at night may be significantly less intense than during the day, the intensity of the system as a whole requires some work to be carried out at night. What is of interest here is that, through the role of cleaning and maintenance, we might see a healthier element of nocturnal urban metabolism. Just as the body refreshes and maintains itself at night, so too does the city. Pushing the metaphor further, we might suggest that this time is essential to the health of urban life. With a lack of down time for sleep, humans can become unwell, as too much strain is placed on the body. Similarly, a lack of down time for the city will stretch resources, prevent maintenance and ultimately wear out the urban environment.

That this work of cleaning and maintenance is often invisible, forgotten and overlooked by those in power can be a problem. Night cleaners across the globe are disproportionately from immigrant or impoverished backgrounds. In many parts of the world they are predominantly women (Brody, 2006), although both Melbin and I have noted that most night-time street cleaners are men. (I was told that the city of Newcastle wants more male cleaners on the street as they are seen as more able to 'handle themselves' among drinkers at night.) In light of her research in Thailand, Brody (2006) describes these workers as the 'cleaners you aren't meant to see'. While clean and modern Western-style shopping malls, offices and other spaces are presented as integral to the Thai vision of modernity, the work of cleaners from ethnic minorities to maintain these spaces is absent from the country's discourse of modernization, reflecting a wider series of absences in development discourses. Research into shift working has identified a series of health problems associated with the disruption to body clocks, reduced exposure to natural light and general fatigue that all night workers, and particularly the low paid, can experience. Here, then, the people at the coal face of the expanding frontier absorb the incessancy of new urban rhythms to the detriment of their health. In turn, this incessancy is passed on to their families, with partners and children of night-shift workers suffering by absorbing the fatigue and difficulties of the night workers themselves. Night work remains an area of employment in which the (urban) poor are disproportionately engaged, and it presents a number of socio-psychological challenges. Thus, the day's dependence on the night for maintenance is also a dependency of the powerful, the wealthy – those who are often able to choose when they are active in a 24/7 society – on the less powerful and less active.

Living with darkness: infrastructure and the possibilities of urban expansion

If we travelled back to a pre-electrification period – say to 1700 – then, globally, we would find some form of nocturnal society in both urban and rural areas. From ships at sea through to ancient rituals, there would be a certain amount of human activity. Society has never stopped completely at night, and artificial illumination has been present for as long as we could meaningfully call humans 'human'. Nonetheless, the emergence of gas and later electric lighting, alongside the creation of safe heating systems in both the home and public spaces, has been and continues to be transformative as far as the urban night is concerned. Artificial lighting allows the earth to appear urban when pictured from above. Yet, this is exactly the sort of bird's-eye view of the planet that ecosophical perspectives (among many others) have attempted to overcome.

In this chapter I have argued that these transformative infrastructures, as well as other systems which hold urban life together, intersect with the natural – or biogeoastronomical – night and wider socio-cultural inequalities in diverse ways. The brightly lit urban frontier quickly breaks down in the face of scrutiny at the scale of the shadows. In turn, this analysis reveals the ways in which the built environment is produced by and produces the everyday behaviour of city inhabitants. While discussions of individual experiences of dark and light are presented in later chapters, the argument so far supports Bateson's (1973) assertion that we must focus on organism *with* environment, not on the two separately.

So what of the promise of an ecosophical urban night? How might we live with the darkness, or with the biogeoastronomical night? In this chapter, I have argued that LED lighting offers the opportunity to make the conditions of possibility of night-time society cleaner, cheaper and more accessible. However, I have also expressed concern at its confluence with smart city technologies, which risk obfuscating decision-making and alienating lighting practice from city and planetary life. Infrastructural cleaning and maintenance must not be overlooked as they are essential to the production of subjectivity and to the reinvigoration of the city. Paying close attention to these peripheries could produce a nocturnal city that is better for all. The three forms of light pollution could all be tackled by the spread of LED lighting, but there are significant challenges to overcome. Most obviously, LEDs have frequently been used to provide brighter lighting at no greater cost, rather than dimmer lighting at a reduced cost. Alongside this consideration of the earth and non-human life, however, we must also be alert to urbanist demands for a lively night-time city. This raises the question of the minimum that is required for the city to persist through the night and remain present in people's lives, even as a borderland. Several answers are possible, but it is clear that the urban night demands a variety of infrastructural practices for this borderland to be maintained; these are the guards, the walls and the condition of the frontier. The governance required for a successful urban world is in part the successful governance and management of these temporal edge-lands.

References

Adkins, E., Eapen, S., Kaluwile, F., Nair, G. and Modi, V. (2010) 'Off-grid energy services for the poor: Introducing LED lighting in the Millennium Villages Project in Malawi', *Energy Policy*, 38(2), pp. 1087–1097.

Alstone, P., Radecsky, K., Jacobson, A. and Mills, E. (2014) 'Field study methods and results from a market trial of LED lighting for night market vendors in rural Kenya', *Light and Engineering*, 22(2), pp. 23–37.

Bailey, R. (2017) *Street Lighting Trend Analysis 2015/16*: http://apse.org.uk/apse/index.cfm/members-area/briefings/2017/17-06-trend-analysis-201516-street-lighting/ (accessed 7 September 2017).

Baldwin, P. C. (2012) *In the Watches of the Night*. Chicago: Unviersity of Chicago Press.

Bateson, G. (1973) *Steps to an Ecology of Mind*. St Albans, Australia: Paladin.
Beaumont, M. (2015) *Night Walking*. London: Verso.
Bogard, P. (2011) 'Why dark skies?', in Bogard, P. (ed.) *Let There Be Night*. Reno: University of Nevada Press, pp. 1–8.
Branch, M.C. (1970) 'Delusions and diffusions of city planning in the United States', *Management Science*, 16(12), pp. B714–B732.
Brenner, N. (2013) 'Theses on urbanization', *Public Culture*, 25(1), pp. 85–114.
Brenner, N. (2014) 'Introduction: Urban theory without an outside', in Brenner, N. (ed.) *Implosions/Explosions*. Berlin: Jovis, pp. 14–31.
Brody, A. (2006) 'The cleaners you aren't meant to see: Order, hygiene and everyday politics in a Bangkok shopping mall', *Antipode*, 38(3), pp. 534–556.
CISCO (2014) *IoE-driven Smart Street Lighting Project Allows Oslo to Reduce Costs, Save Energy, Provide Better Service*: www.cisco.com/c/dam/m/en_us/ioe/public_sector/pdfs/jurisdictions/Oslo_Jurisdiction_Profile_051214REV.pdf (accessed 8 September 2017).
Cornell Jr, J.C. (1971) 'Light pollution from a growing Tucson threatens "astronomy capital of the world"', *New York Times*, 20 June: http://search.proquest.com/docview/119343548?accountid=12753. (accessed 19 October 2017)
Ekrich, A.R. (2006) *At Day's Close*. London: Pheonix.
Foucault, M. (1963) *The Birth of the Clinic*. London: Routledge.
Foucault, M. (1970) *The Order of Things: An Archaeology of the Human Sciences*. London: Routledge.
Foucault, M. (1979) *Discipline and Punish: The Birth of the Prison*. Harmondsworth: Penguin.
Gandy, M. (2004) 'Rethinking urban metabolism: Water, space and the modern city', *City*, 8(3), pp. 363–379.
Gaston, K.J., Davies, T.W., Bennie, J. and Hopkins, J. (2012) 'Review: Reducing the ecological consequences of night-time light pollution: Options and developments', *Journal of Applied Ecology*, 49(6), pp. 1256–1266.
Goffman, E. (1974) *Frame Analysis*. London: Penguin.
Graham, S. and Thrift, N. (2007) 'Out of order: Understanding repair and maintenance', *Theory, Culture and Society*, 24(3), pp. 1–25.
Green, J., Perkins, C., Steinbach, R. and Edwards, P. (2015) 'Reduced street lighting at night and health: A rapid appraisal of public views in England and Wales', *Health and Place*, 34, pp. 171–180.
Hölker, F., Moss, T., Griefahn, B., Kloas, W., Voigt, C.C., Henckel, D., Hänel, A., Kappeler, P.M., Völker, S., Schwope, A., Franke, S., Uhrlandt, D., Fischer, J., Klenke, R., Wolter, C. and Tockner, K. (2009) 'The dark side of light: A transdisciplinary research agenda for light pollution policy', *Ecology and Society*, 15(4), p. 13.
Isenstadt, S., Maile-Petty, M. and Neumann, D. (eds) (2014) *Cities of Light: Two Centuries of Urban Illumination*. London: Routledge.
Jakle, J.A. (2001) *City Lights*. Baltimore: Johns Hopkins University Press.
Knauer, R.S. (1980) 'Light suppresses melatonm secretion in humans', *Science*, 210, p. 12.
Kumar, A. (2015) 'Energy access in an era of low carbon transitions: Politicising energy for development projects in India'. Unpublished Ph.D. thesis, Durham University.
Lemaire, X. (2011) 'Off-grid electrification with solar home systems: The experience of a fee-for-service concession in South Africa', *Energy for Sustainable Development*, 15(3), pp. 277–283.

Longcore, T. and Rich, C. (2004) 'Ecological light pollution', *Frontiers in Ecology and the Environment*, 2(4), pp. 191–198.
Melbin, M. (1987) *Night as Frontier: Colonizing the World after Dark*. New York: The Free Press.
Morton, T. (2007) *Ecology without Nature*. Cambridge, MA: Harvard University Press.
Navara, K.J. and Nelson, R.J. (2007) 'The dark side of light at night: Physiological, epidemiological, and ecological consequences', *Journal of Pineal Research*, 43(3), pp. 215–224.
Norman, W. (2011) *Rough Nights: The Growing Dangers of Working at Night*: https://youngfoundation.org/wp-content/uploads/2012/10/Rough_Nights.pdf (accessed 8 September 2017).
Nye, D. (2010) *When the Lights Went out: A History of Blackouts in America*. Cambridge, MA: MIT Press.
Nye, D. (2014) 'Foreword', in Isenstadt, S., Maile-Petty, M. and Neumann, D. (eds) *Cities of Light: Two Centuries of Urban Illumination*. London: Routledge, pp. xxi–xxiii.
Osayimwese, I. (2014) 'Lagos', in Isenstadt, S., Maile-Petty, M. and Neumann, D. (eds) *Cities of Light: Two Centuries of Urban Illumination*. London: Routledge, pp. 79–81.
Park, O. (1940) 'Nocturnalism: The development of a problem', *Ecological Monographs*, 10(3), pp. 485–536.
Pauley, S.M. (2004) 'Lighting for the human circadian clock: Recent research indicates that lighting has become a public health issue', *Medical Hypotheses*, 63(4), pp. 588–596.
Pode, R. (2010) 'Solution to enhance the acceptability of solar-powered LED lighting technology', *Renewable and Sustainable Energy Reviews*, 14(3), pp. 1096–1103.
Pode, R. (2013) 'Financing LED solar home systems in developing countries', *Renewable and Sustainable Energy Reviews*, 25, pp. 596–629.
Powells, G., Bulkeley, H., Bell, S. and Judson, E. (2014) 'Peak electricity demand and the flexibility of everyday life', Geoforum, 55, pp. 43–52.
Pritchard, S.B. (2017) 'The trouble with darkness: NASA's Suomi satellite images of earth at night', Environmental History, online early access: https://academic.oup.com/envhis/article-abstract/22/2/312/2998686/The-Trouble-with-Darkness-NASA-s-Suomi-Satellite (accessed 19 October 2017).
Riegel, K.W. (1973) 'Light pollution', *Science*, 179(4080), pp. 1285–1291.
Schivelbusch, W. (1988) *Disenchanted Night*. Oxford: Berg.
Schlör, J. (1998) *Nights in the Big City: Paris, Berlin, London 1840–1930*. London: Reaktion.
Smith, M. (2009) 'Time to turn off the lights', *Nature*, 457(7225), p. 27.
Steinbach, R., Perkins, C., Tompson, L., Johnson, S., Armstrong, B., Green, J., Grundy, C., Wilkinson, P. and Edwards, P. (2015) 'The effect of reduced street lighting on road casualties and crime in England and Wales: Controlled interrupted time series analysis', *Journal of Epidemiology and Community Health*, 69(11), pp. 1118–1124.

4 Nightlife and night-time economy

Whether in the darkness of a winter night in temperate latitudes, or the cooling sunset of a lower-latitude evening, the distinction between urban and rural ways of living perhaps comes out most powerfully at night. Rural 'nightlife' is far from absent, but in many locations it consists of a small number of centralized events: single bars or cafés, or gatherings of a few people in one or two homes. This pales in comparison with the party districts of cities with night*lives*; with night markets and night-time shopping malls; with the industrial estates, offices and factories that continue to operate 24/7 in many parts of the world; with the busy hum of airports or other global infrastructural sites. If the previous chapter looked at the minimum necessary infrastructure for the night-time city, this chapter is interested in what happens when the city doubles down and shows its most vibrant night-time form. In other words, this is about 'the night-time economy' or, more accurately, the hospitality and tourist industries at night that offer leisure, partying and relaxation.

Although nightlife is perhaps the most easily identifiable aspect of the nocturnal city, it constitutes only one part of the actual night-time city. How can we bring the extensive research into this area to the question of planetary urbanization? Urban nightlife – and the 'night-time economy', as it has been labelled – in many ways reveals how partial and limited the spread of activity actually is in the contemporary night-time city. Starting with a case study of the night-time economy in the UK, this chapter explores the ways in which – notwithstanding considerable growth and despite being understood as vibrant, lively and dynamic – almost every city remains reduced, limited and significantly less active at night than during the day. In the second half of the chapter I will explore a wider range of urban night-time leisure activities. The aim is to delve into the night-time city in its most vibrant, most densely populated, most day-like state and explore what this has to tell us about the edge-lands of planetary urbanism, while also showing how thin the edge truly is.

Booze, bingeing and beer: the night-time economy in the UK

Living for the weekend; 'the night-out'; weekend hedonism; 'a weekend wasted is never a wasted weekend'. Whatever the slogan, the description, the evocation, the ideal of a Friday or Saturday night spent drinking alcohol in the city centre is pervasive in the UK, and, with a greater or lesser focus on alcohol, in many other countries too. Anticipated throughout the week, the night out and the night-time economy drive many perceptions of what the city centre at night is and what it can do (Roberts and Eldridge, 2009; Jayne et al., 2011). This night-time economy is associated with desire, in the form of both sexual passion and a psycho-social yearning for release from the rhythms, routines and controls of the working week. In the UK, these intersect in a conflation of alcohol, sex and possibility, driven by potential futures of gaining something new, often love in its many forms: sex and lust, voyeurism, flirting (Radley, 2003) or meeting a long-term partner (Winlow and Hall, 2009). The mantra is: 'Never mind what has happened previously, *tonight* will be the night for fun.'

This, though, tells only a small part of the story. Even in the UK, the typical night out consists of more than just drinking, sex and dancing. Smoking, chatting, playing slot machines, eating, queuing, taxi or public transport journeys, walking, arguing and more all form integral parts of the experience of the city centre at night. Academia has been relatively slow to investigate these elements, and particularly slow at looking at non-drinking practices in relation to the night-time economy (Roberts and Eldridge, 2009). A reflection on 'para-drinking' activities presents a wider range of participants in the night-time city, counteracting the rather homogeneous depiction that has dominated popular discourse and debate (Jayne et al., 2011). This follows efforts by a range of researchers to widen the scope of the people and objects that are viewed as integral parts of the night-time economy (Jayne et al., 2012; Wilkinson, 2015). A focus on para-drinking activities may prove useful when trying to differentiate between diverse groups of drinkers or drinking experiences: a night in which the fruit machine plays a significant role can be very different from a night spent queuing outside an exclusive club (Jackson, 2004), or one spent talking quietly with friends over a bottle of wine.

The UK's night-time economy is simultaneously a singular, dominant force which shapes the country's night-time cities and a multiple, amorphous assemblage of practices, spaces, powers, affects, emotions and experiences. As such, it is perhaps unsurprising to find that the timespaces of the night-time economy are highly contested. If we turn to the emergence of night-time entertainment cultures in the eighteenth century, we can see that this contestation over how people spend their time in the city at night, and with whom they associate, is far from new (Schlör, 1998). Through the nineteenth century, issues of the gender and identity of drinkers became highly pertinent, and these persist (Beckingham, 2012). Gwiazdzinski (2005) speaks of a night which will be defined by the future of these conflicts, and whether the leisure users of cities are absorbed into a night-time version of the daytime high street (as in much of the UK), or whether the hedonism, alterity and counterculture of the night continue despite the advance of capitalism. On the one hand, a pervasive image of the night as a timespace of counterculture still prevails.

People understand drinkers as dangerous, outside of normal legal and moral practices. Revellers go out at night in search of experiences that they could not encounter in the sober city of day. On the other hand, the rhythm of relaxation and hedonism, in relation to work, and the affects of anticipation that I described earlier may be a form of 'cruel optimism' (Berlant, 2011). In other words, the presence of the night-time economy offers as pyscho-social reassurance of opportunities for difference and release which allows people to work through unequal, alienating and unfair conditions. With this claim, the night-time economy might contribute to people's acceptance of difficult, mundane or precarious work by offering potential, if temporary, respite. In practice, the night-time economy is likely to have both positive and negative personal outcomes, producing subjectivities in contradictory ways.

Creating nightlife: the emergence of the British night-time economy

The night-time alcohol and leisure industry in the UK serves as a good case study as to how capitalist expansion in the night has altered urban life. In 1991, a Comedia report titled *Out of Hours* declared that 'there was little urban life to be involved with' at night in many of the towns and cities that the researchers visited (Comedia, 1991, p. 5). At that moment in time, Comedia – a think-tank concerned with shaping urban planning in the UK – identified a key opportunity to change the future of the assemblage known as the 'night-time economy'. The organization, and aligned writers such as Bianchini (1995), worked hard to name this assemblage *the* night-time economy, a phrase which singularizes the multiplicity of the urban night as a single thing and prioritizes its economic elements over socio-cultural harms or benefits. What was being sold was thus 'the opportunity of "doubling" the city's economy, starting perhaps from entertainment but then widening into other areas' (p. 124). I much prefer the term 'late night alcohol and leisure industry' when referring to the bars, pubs and clubs of the UK at night, though I can fully understand why this significantly less catchy phrase did not achieve the cachet of 'the night-time economy'. Nonetheless, it is important to note that the term 'night-time economy' conceals the fact that people's activities in bars, clubs and pubs comprise just one aspect of any city's nocturnal economic activity and form just one of the social, political, economic and cultural prompts that collectively shape contemporary urban night-time leisure spaces.

Broadly, four major coinciding trends created the timespace in which the UK's late night alcohol and leisure industry could develop. From the 1980s onwards, the country's legislation relating to alcohol consumption was modified in line with emerging neoliberal policies (Peck and Tickell, 2002) . In particular, in 1989, a government report recommended breaking up what was effectively a cartel of six breweries. These companies dominated the industry through the 'tied' system of pub ownership, whereby landlords ran their establishments as tenants and were obliged to purchase beer and other drinks from the parent company. The new legislation forced the 'big six' to dispose of thousands of pubs, and to allow their tenants to negotiate their own drink-purchasing deals. 'Pubcos' (pub companies) emerged, taking advantage of the glut of available pubs to create a number of pub and bar chains. They also made full use of regeneration projects which were starting to have a positive impact on the country's city centres after decades of decline. While most of the early projects had been business-property led, developers increasingly saw the value of incorporating facilities such as bars, clubs, restaurants and pubs that would encourage people to spend a few hours in the inner city at the end of the working day, and might even persuade them to live in city centres rather than commute from the suburbs. Pubcos thus established themselves as active participants in the construction of new nightlife developments and the acquisition of large, empty public buildings, such as old shops, banks, town halls, cinemas and factories, moving in just as older services and businesses were vacating city centres. These developments suited the local authorities, who were happy to see large buildings reoccupied.

While these economic and political forces were at play, a series of socio-cultural factors were also important. The alcohol industry suffered declining sales between 1985 and 1995, particularly among young people, many of whom preferred ecstasy, as part of dance and rave culture. In the face of this competition, the alcohol industry was forced to rethink its marketing. A combination of new products and pub redesigns attempted to attract a younger and more diverse clientele to drinking establishments. In particular, pre-mixed 'alcopops' were designed to be sweet, cheap and potent, targeting in particular women (who, in contrast to men, were a burgeoning market for the drinks industry at the time), younger drinkers and people who wanted drinks that would fit with dancing. Redesigns also made pubs more female friendly. Many women viewed traditional British pubs as intimidating spaces, due to their small rooms and the presence of a largely male, ritualized culture. Pub owners attempted to reduce this intimidation and attract more women by converting 'dingy' rooms into larger and brighter spaces. Finally, the UK's late night alcohol and leisure industry benefited from a period of sustained economic growth in Western Europe from the mid-1990s until the economic crisis of the late 2000s. During this time, a great deal of surplus capital was invested in city centres (Harvey, 2012), and pubs and bars offered excellent investment opportunities as they enjoyed steady income from the business of selling alcohol as well as exponential growth in the value of the land on which they were built. Indeed, many of the new pubcos were, in effect, property development companies which viewed the alcohol-selling and leisure aspects of their business merely as a means of generating a dividend on their investment.

So the emergence of 'night-time high streets' in the UK was driven by the intersection of a number of social, cultural and economic changes, with the outcome being the production of new forms of being in the city: that is, new subjectivities centred on urban living and hedonistic weekends. The nightlife that was constructed in this boom period by and large still exists today, although the global economic crisis saw some contraction in the market, and more recently the number of pubs and nightclubs has started to decline. Nonetheless, pubs, the traditional form of late night venue in the UK, remain popular in cities, towns, suburbs and villages. However, they have become more diverse. In addition to traditional 'boozers', which continue to focus on (cheap) beer and a largely white, male clientele, the UK boasts: 'gastropubs', which put food retail at the core of their business; themed chain pubs that seek to attract large numbers of people through a variety of initiatives, such as cut-price alcohol deals or family-friendly play areas; 'locals', which are as varied as the neighbourhoods in which they are situated; cocktail or gin bars, which focus on specific drinks; 'craft ale' pubs that sell a range of beers from small-scale producers; and numerous other venues which attempt to exploit a variety of niches. On a typical weekend evening, people often follow well-trodden paths and visit several venues, as well as making use of taxis or public transport. In urban areas, drinkers venture into the small fragments of the town or city that remain open after dark, especially the night-time high streets that were developed in the peak period of expansion in this sector, where we now find clusters of late night pubs, bars and clubs. Larger cities have a more diverse range of offerings, with music venues, comedy clubs, cafés, independent cinemas and casinos, though the choice declines rapidly outside of the ten to fifteen most populous UK cities. Closing times remain between 02.00 and 03.00 for the majority of venues, though a small number of clubs may continue until 04.00, particularly at the weekend; later openings are rare outside London. Throughout the night, people make purchases from fast-food venues. These are still mainly the chip, kebab and pizza takeaways that have dominated Britain's late night food business since the early 1990s, though there has been some diversification, with places selling falafel and noodles, as well as some chains, such as McDonald's and Greggs, introducing all-night opening.

Since this period of rapid growth, the picture of the night-time alcohol and leisure industry has largely been one of shifting trends, often in line with economic changes, rather than another major reorganization. In the popular imagination the public urban night thus remains a carnivalesque timespace that young men and women enter to see and be seen, and into which they place their hopes and expectations for a time-limited but exciting period of play: what Szmigin *et al.* (2008) have labelled 'calculated hedonism'. Because of this, a significant moral and political debate has emerged. From the reports of certain newspapers (the *Daily Mail* has included an article on the drunken 'debauchery' of New Year's Eve revellers every 1 January since 2007), it is easy to believe that public space at night is a hotbed of violence and danger. Furthermore, and broadly defined, 'harm' from alcohol – whether in the form of liver disease, financial difficulties or violent crime – rose across the UK in parallel with the growth in alcohol consumption from the early 1990s to the mid-2000s (Jayne *et al.*, 2011).

While this portrayal does capture a commonly experienced night out in many British cities, it is limited. As well as the ways in which the night-time alcohol and leisure industry has continued to evolve, the urban night extends far beyond this description. What is labelled the UK's 'night-time economy' in fact looks much more like what Gwiazdzinski (2005) calls the 'archipelagos' of the night: islands of activity that are connected by lines of taxis, walkers and night buses. Even in large cities with more complex and multifaceted urban nights this late night alcohol and leisure industry forms only part of the economic activity of the city at night. Furthermore, researchers in this area have shown that late night drinking is much more complex and diverse than this picture reveals: it is practised by people of different ages, ethnicities and religions, and occurs not just in pubs but also in private homes, public squares and parks (Jayne et al., 2011; Stepney, 2014; Bøhling, 2015; Wilkinson, 2015).

Being nocturnal: night-time subjectivities

So, from the early 1990s until the economic crisis of the late 2000s, the UK's towns and cities underwent a steady nocturnal transformation, the aftermath of which persists. This transformation contributed to the production of new forms of urban nocturnal subjectivity. To reiterate the argument made in Chapter 1, subjectivities in cities seem to emerge out of the circulation of affects, discourses and bodies. To put this another way, this description of the urban night as a site of the production of subjectivity is an understanding of the urban night as a timespace in which subjects are made. As Guattari (1993, pp. 143–144) argues,

> the span of developed/constructed spaces extends quite beyond their visible and functional structures. They are essentially machines, machines of meaning, of sensation, abstract machines ... that carry incorporeal universes that are not universals but that can standardize individual and collective subjectivity.

In the opening chapter of this book, I described the city as a subjectivity machine, and the night-time alcohol and leisure industry makes these processes particularly visible. Urban planners and bar owners have sought to populate the night-time city with the 'right' kind of subjects: people who are keen to consume and spend money on food, alcohol and dancing, without resorting to violence or succumbing to drunkenness. Looking at the contemporary night-time high streets of the UK we can see that, despite a few shifts since the period of peak growth, several features have coalesced to create the conditions for this production of subjectivity.

It is useful to divide the tools of the night-time subjectivity machine into four different components. First, affects circulate, pushing and shaping bodies in a variety of ways. They are produced out of all of the stimuli that circulate and shape the self, received and interpreted in the process of – or even before – conscious recognition (Deleuze, 1978). In the urban night, bodies are more open to affect: first, as the reduced numbers of people make those interactions which do occur more intense; and, second, as the interplay of light and dark reduces the strength of visually determined boundaries around the self. Several practices in the night, such as the use of light, sound and alcohol, enhance this process, shaping the ways in which participants in the urban night interact with one another. This can be illustrated by the transformations that take place in the Wetherspoons pub chain as daytime turns into night. During the day, these pubs are oriented towards the sale of cheap food, with coffees, meals and snacks forming a core part of the business. As evening falls, however, they become much more oriented towards the sale of alcohol and, in some venues, dancing. In part, the change that occurs in the space is achieved through a change in various forms of affective manipulation. Inside the pub, the lighting is dimmed (see Bille et al., 2015 on the affectivity of lighting), the music volume is increased and some of the more comfortable seating is removed. These changes influence how people interact with one another – they lean closer, shout louder but talk less, drink more quickly and/ or start dancing. These interactions are all changes in the ways in which bodies affect one another and are affected by others, and they alter the emergent subjectivities of the people who are affecting one another.

Second, the urban night produces subjectivities through discourses and the circulation of meaning. As Hollands (1995) argues, the weekend 'big night out' still plays an important role in many people's identities. In the context of the decline of traditional industrial lifestyles, the traditional Friday or Saturday night out connects young people with the past: in many towns and cities, they patronize the same venues and pubs that their parents and even grandparents visited decades earlier (Chatterton and Hollands, 2001). Among young people, the urban night is a space for making the self through rebellious experiences and risk-taking – important rituals in the process of becoming adult (Stepney, 2014). These experiences are key within numerous youth subcultures – from the hedonism of the clubbing scene (Malbon, 1999) to the activities of goths, punks and others who attend music gigs, the night-time alcohol and leisure industry can be a cornerstone of identity. Subjectivities in the urban night are also produced through the power of demonization, with drinkers described or labelled as (variously) violent, dangerous and threatening. Researchers have repeatedly shown that drunken behaviour is culturally learned (Critchlow, 1986) and that discourses of unruly drinking contribute to this learning.

As well as this immaterial framing, however, the urban night is characterized by a series of material or infrastructural manipulations of subjectivity. Returning to Guattari (1993), it should be noted that his example of the plurality of urban subjectivity comes from an experience of memories and emotions brought to mind by crossing a bridge in São Paolo. This illustrates in particular the role of the *built* urban environment, which contains cues that shape our everyday experience. As such, these acts of planning, governing and management of the night-time city build the subjectivities that are emergent from the city. People queue on streets, surrounded by police officers or door staff who act as beacons of surveillance and official security. Alleyways become spaces for drinking, urinating, sex or other activities that require avoidance of that surveillance. Alcohol itself flows around the city centre, whether drunk on benches or in public squares, spilled onto streets or carried within bodies. Urban spaces which in the day are convivial and benign can at night be threatening or dangerous, while others which are dull and quiet can become lively and vibrant. All of this shapes the people who experience these spaces in new ways.

These different prompts – affects, discourses and materiality – combine to produce a public urban night that is best understood as 'atmosphere'. Atmospheres are assemblages that have gained control over a particular timespace; they have what de Certeau (1984) calls their 'proper place', using what in French is word-play over the dual meaning of the word *'propre'* as connoting both 'proper' (correct) and 'own' (belonging to). They combine the elements outlined above, intersect with existing subjectivities and encourage new ones to emerge. It is for this reason that an atmosphere may be attractive to some and scary for others; identified by some and missed by others; present on some nights, but absent on others. In other words, an atmosphere will attract certain people more than others, and in the case of the UK's night-time economy this has often worked to reinforce existing inequalities in access to and freedom to use public space. As Hobbs *et al.* (2003, p. 273) note, in the urban night, 'it is those most thoroughly seduced of consumers, to the tune of a dozen lagers, who are most inclined to be targeted by swarming police units, teams of bouncers and couplets of street wardens'. Only some of the users of the night-time city are swept along by this atmosphere; others miss it, or reject it, or are simply less taken by it.

Understanding the urban night that is produced by the night-time alcohol and leisure industry in the UK as an atmosphere prompts wider discussions of the urban public night and nightlife. We might say that night has fragmented, with a solely nocturnal atmosphere giving way, at least in part, to one shaped by day-like features. Alternatively, this process might be understood as a complicating of the nocturnal atmosphere, possibly as a wider range of atmospheres is produced. In the context of the UK's night-time alcohol and leisure industry, the fractured night-time atmosphere is intense due to the consumption of alcohol, the psycho-social value that is placed on the night as a time for relaxation and release, and the manipulation of night-time city spaces, deliberately or otherwise, to encourage the emergence of certain subjectivities.

So the night-time economy of British cities serves as a key example of the ways in which changes to the urban night, driven by capitalist expansion, have reshaped both urban space and subjectivities. New night-time high streets and new ways of inhabiting the city have been produced. Turning next to the global scale, I want to explore the ways in which nightlife has spread in a variety of contexts. The following examples do not represent an exhaustive study of all the world's night-time economies, but they do reveal some of the diversity in this part of the night-time frontier.

Global nocturnal leisure

Taipei: night markets

Night markets have a long history in Chinese culture, with evidence for them dating back at least to the Tang Dynasty in the eighth and ninth centuries CE. They subsequently spread throughout South-East and East Asia, where they have typically been areas of both commerce and socialization: relaxed, informal meeting-places where people shop but also interact, observe, promenade and take advantage of the public space (Yu, 2004). Lit by candles and oil lamps, these sites historically sold both the products that one might find during day and 'night market special goods', which were typically snack foods and fortune-telling services (p. 134), reflecting an increased focus on leisure, spirituality and hedonism that we see in nights elsewhere. The products of the night market thus reveal its role in releasing some of the pressures of the day; this was a more relaxed offering than was available in the daytime. They produced intensified spaces of social interaction and subjectivity production; indeed, Yu describes them as 'carnivalesque' (note the similarity here to the UK example, above). Today night markets are most associated with Taiwan, although others, known as '*pasar malam*', persist in Indonesia, Malaysia and Singapore (Ibrahim and Leng, 2003). In Taipei, they boomed during the 1960s and 1970s as a way of selling surplus products from Taiwan's growing manufacturing industry alongside the familiar snacks of the traditional Chinese night market (Yu, 2004), and they have continued to grow in popularity, attracting tourists and becoming part of contemporary Taiwanese identity (Lin et al., 2011). In some places they have been absorbed by mainstream and state activity, with official night markets operated by the government.

Taipei's largest night market is located at Shilin. Opening from 16.00 until around midnight, it is firmly on Taipei's tourist trail: at the time of writing, it was third on TripAdvisor's list of the city's shopping experiences (which was topped by the smaller Raohe night market). Shilin features outdoor stalls, an indoor food court and a prevalence of neon lighting. This lighting forms a key part of the sensual pleasure of the market, but it contrasts with the dim street-level lamps and oil lanterns, which many stall owners still use. Alcohol, while present, plays a minor role, yet this is nonetheless a space for partying, relaxation, sociability and subjectivity manipulation. This reveals the ways in which the interplay of affects and darkness forms a key aspect of the atmosphere of the night-time city. In other words, while drinking clearly plays a central role in the UK's night-time alcohol and leisure industry, the night markets show that the conviviality of cities at night often rests on more than mere alcohol consumption. Taipei's markets are a twentieth-century reinvention of traditional East Asian nocturnal shopping culture, just as the UK's night-time economy is a reinvention of traditional nocturnal public house culture.

Interestingly, Taipei's night markets and the UK's night-time economy are currently facing similar criticism. The Leuha market was threatened with closure by the state in 2015 over licensing regulations, but the case was more broadly seen as a response to the illegal trading and noise that are routinely associated with such places. Although they are artificially lit hubs of activity, at night people can quickly slip into the shadows, so the night markets provide opportunities for sellers of illicit goods and others to operate outside of state control. Therefore, despite their appeal as tourist attractions, and their central role in Chinese culture, the government views them with some suspicion (Yu, 2004). Thus, once again, we see an atmosphere of alterity, liminality and 'riskiness' associated with night.

In recent years, East Asia's night markets have been repackaged for Western consumers, and versions of them now even tour around the world. This is indicative of the increased worldliness of the urban night, with, as in other fields, ideas spreading much more readily between distant places.

Sydney: the state fights back

But what happens when the state attempts to establish control over the hedonism of the night? Sydney has long been famous for its active nightlife. For instance, Montgomery (1995), in an article designed to encourage British cities to expand their night-time economies, cites the city as a great example of vibrant night-time culture. However, Sydney has recently undergone something of a nocturnal transformation. 'Lockout Laws' were introduced in January 2014 in response to the deaths of two men in 'one-punch' night-time attacks in the city's central business district. These laws made two major changes to Sydney's nightlife. First, all alcohol licences which extended beyond 03.00 were revoked. Second, and much more damagingly for the night-time alcohol and leisure industry, customers were prohibited from entering any bar after 01.30. The intention was twofold: first, to stop people who had been ejected from one bar from simply going to another; and, second, to reduce the crowds on the streets by eliminating the possibility of bar crawls in the early hours of the morning. In other words, the main aim was to tackle street violence by dissuading people from walking through the streets. Now, if they want to have a night out, they have to remain *inside* a venue. These laws have provided an illuminating insight into the symbiosis of public and private space in Sydney at night, with the legislation specifically designed to regulate not only the bars and clubs but also the streets outside their doors.

The response has been mixed. Public health campaigners have called for the introduction of similar laws throughout Australia, and the Australian Bureau of Crime Statistics and Research recorded a 32 per cent drop in assaults in the key Kings Cross district of Sydney, with assaults across the wider CBD falling by 26 per cent. Moreover, an independent review of the laws supported these findings by noting that there was no evidence that the violence had simply been displaced elsewhere, and that alcohol-related domestic violence had also fallen (Callian, 2016). On the other hand, more recent figures for 2017 do suggest a small increase in street violence in some suburban locations, possibly due to displacement from the city centre (Donnelly et al., 2017), while both the night-time alcohol and leisure industry and many of the people who previously used the city at night have protested strongly against the laws. Clubs and bars reported a 20 per cent downturn in clientele within weeks of the laws' introduction. Although it is difficult to find exact figures for closures, just over two years after the laws had passed the *Sydney Morning Herald* published an article titled 'Going, going, gone: 10 iconic bar closures and moves in Sydney' (Dumas, 2016). Furthermore, the closure of late night venues has had a knock-on effect: the bars where people used to meet in the early evening and the restaurants where they used to eat have all suffered as much as the nightclubs themselves. As the journalist Matt Barrie (2016) wrote in a widely shared blog entry, 'the total and utter destruction of Sydney's nightlife is almost complete'. The aforementioned independent review, despite backing the laws, also found that 'Pedestrian traffic is greatly reduced in the [central] Precincts. Some live entertainers have lost their employment or are reduced in their employment as live entertainers ... opportunities for employment in the Precincts have been reduced' (Callian, 2016, pp. 4–5). In 2016 these closures and job losses prompted the founding of the 'Keep Sydney Open' campaign to protest against the laws. This has already attracted over 15,000 supporters, including venue staff, consumers and music producers. At the time of writing in mid-2017, the laws were still in place, although the lockout times have been relaxed for live music venues in a bid to avoid further closures and job losses.

Sydney reflects what is at stake in the management and control of public activity at night. Callian (2006, p. 4) neatly summarizes the challenge presented by the city's night-time economy: 'Vibrancy was unfortunately, however, accompanied, probably as a function of the numbers and density of visitors to the Precincts, by a degree of antisocial, and in the residents' terms, squalid and sleazy behaviour.' On the one hand, the night's atmosphere of conviviality, riskiness and hedonism forms a key part of many people's lifestyles, and it attracts tourists as part of a vibrant and lively city. The presence of such a timespace is particularly important for Sydney, a city which promotes itself as young, lively and free. It also creates employment and contributes to the city's liveability. However, the vibrancy is counterbalanced by the fact that there are undoubted public health benefits to placing restrictions on the night-time alcohol and leisure industry: lower levels of violence, less antisocial behaviour and reductions in long-term physical harm caused by alcohol abuse. Sydney's conundrum is one that is shared by all frontiers (Melbin, 1987): how to manage a space in which the essential qualities of encounter and mixing may threaten the accessibility of the place itself, or the well-being of those who use it.

Kampala karaoke

Karaoke is an unusual phenomenon. Born in 1970s Japan, it has spread globally to be a mainstay of the night-time alcohol and leisure industry in many cities. Wasswa-Matovu (2012) reports on its particularly interesting rise in Kampala, Uganda. Here, the word 'karaoke' is now used to describe a wide variety of night-time vernacular performance, including dance and comedy, though singing along to recordings of popular songs is still at its core. Wasswa-Matovu argues that Kampala's karaoke, in all its forms, provides significant employment opportunities for young people as members of bands, bar staff and entertainment troupes in the city. However, this is a dangerous world, presenting some of the same challenges with regards to violence and sexual exploitation that the night-time alcohol and leisure industry faces in other countries. The presence of karaoke in Kampala reveals a nightlife that is simultaneously local and global, illustrating the city's multi-continental adaptation and the sort of hybridity that is found in many of Africa's capitals. In particular, the pastime's popularity has been attributed to the freedom of expression it afforded the Ugandan people in the post-Amin years: 'Young people are allowed to "become" those images that are out there in the world known to them only through tape and video' (Ssewakiryanga and Isabirye, 2006, p. 67). Kampala's karaoke scene has thus emerged from a 'grassroots' source, representing a vanguard of culture rather than the mainstream of the city's night-time alcohol and leisure industry (Wasswa-Matovu, 2012).

Therefore, karaoke encourages the emergence of atmospheres of defiance and rebellion in a country in which freedom of association and expression are relatively recently won rights. Kampala's nightlife scene is now sufficiently well established to appear on the tourist radar. However, it is a risky timespace. The main concern is over prostitution, and the associated spread of HIV. Both of these issues are closely associated with the karaoke scene, with some bar owners acting as pimps and coercing the performers into sex work (Schulkind et al., 2016, p. 775). The prevalence of prostitution in Kampala at night sits uneasily in a highly religious nation, where fundamentalist Christian values have been incorporated into state policy: for example, in 2013, laws were passed to recriminalize homosexuality, and sentences have been harsh. Sex work, however, is common in the capital at night and forms a key obstacle against many development projects (Mbonye et al., 2012).

Once again, then, we see a night-time city which shimmers on a liminal barrier between an animated nightlife that forms a key part of the urban system and a space of danger, violence and inequality. In Kampala, a Japanese form of night-time entertainment has travelled thousands of miles to mix with a postcolonial African night-time city and produce a risky yet liberating timespace of entertainment.

Contact zones of nightlife

These three examples – Taipei, Sydney and Kampala – as well as the story of the UK's night-time economy show that cities 'come to life' at night in multifarious ways. All of these places host forms of 'risky' behaviour that threaten – or have been deemed to threaten – a number of moral, social, political or ethical guardians. Thus, in a variety of locations, urban nightlife straddles the periphery and the mainstream. In other words, it looks rather like a space which is on the boundary of capitalist expansion, as illegal, hedonistic or dangerous practices come into contact with the publicly open city of the daytime. The night, in this sense, can be understood as a frontier in the battle between dominant urban forces and those on the edge-lands of urban life. In this chapter, I have shown how the night-time alcohol and leisure industry can be understood as emerging out of a particularly controlled and managed atmosphere. The presence of alcohol, the change in social expectations and the intensity produced by the interplay of light and dark combine to create a night-time city which is more malleable than its daytime counterpart, one in which governance is enacted by various groups who seek to attract the 'right' sort of night-time city users so that tendencies towards difference are excluded. At the same time, a range of other people enter the city at night in search of lifestyles that celebrate escape and hedonism.

On the one hand, researchers have repeatedly shown the ways in which the public city at night is a timespace that has been closed off, controlled and subjugated. On the other, as society has become ever more complex and daytime society has 'advanced', the night has been subsumed into what Crary (2013) calls a '24/7 system' – a process that Lefebvre (2004) views as night becoming part of day. Here, the concept of 'contact zones', introduced in Chapter 2, might again be useful. The contact zone describes how difference and in particular the centre and periphery come together at frontiers. Contrasting ways of living clash with each other, with neither fully able to control or lay claim to the space (Morrissey, 2005). The result is that the night-time city hosts vibrant sites in which with the mainstream is constantly adapted by those who seek a night of hedonism and leisure. It is this vibrancy which the well-meaning restrictions of a city such as Sydney endanger.

The contact zone concept also helps to explain why the expansion of capital into the night may be viewed as 'fragmentation': it does not come to dominate the night-time city to the same extent as it does the day; rather, it splits and diversifies through processes of contact. In this way, night might act as the periphery to the urban world – within that world but still acting as a border or edge-land to its more dominant and powerful forces.

References

Barrie, M. (2016) Would the last person in Sydney please turn the lights out? LinkedIn Pulse: www.linkedin.com/pulse/would-last-person-sydney-please-turn-lights-out-matt-barrie (accessed 23 October 2017).

Beckingham, D. (2012) 'Gender, space, and drunkenness: Liverpool's licensed premises, 1860–1914', *Annals of the Association of American Geographers*, 102(3), pp. 647–666.

Berlant, L. (2011) *Cruel Optimism*. Durham, NC: Duke University Press.

Bianchini, F. (1995) 'Night cultures, night economies', *Planning Practice and Research*, 10(2), pp. 121–126.
Bille, M., Bjerregaard, P. and Sørensen, T.F. (2015) 'Staging atmospheres: Materiality, culture, and the texture of the in-between', *Emotion, Space and Society*, 15, pp. 31–38.
Bøhling, F. (2015) 'Alcoholic assemblages: Exploring fluid subjects in the night-time economy', *Geoforum*, 58, pp. 132–142.
Callian, I. (2016) *Review of Ammendments to the Liqor Act 2007 (NSW)*. Sydney: New South Wales Government: www.liquorandgaming.nsw.gov.au/Documents/public-con sultation/independent%20liquor%20law%20review/Liquor-Law-Review-Report.pdf (accessed 30 May 2017).
Chatterton, P. and Hollands, R. (2001) *Changing Our 'Toon': Youth, Nightlife and Urban Change in Newcastle*. Newcastle upon Tyne: University of Newcastle upon Tyne.
Comedia (1991) *Out of Hours: A Study of Economic, Social and Cultural Life in Twelve Town Centres in the UK*. London: Comedia.
Crary, J. (2013) *24/7*. Los Angeles: Verso.
Critchlow, B. (1986) 'The powers of John Barleycorn: Beliefs about the effects of alcohol on social behavior', *American Psychologist*, 41(7), pp. 751–764.
de Certeau, M. (1984) *The Practice of Everyday Life*. Berkeley: University of California Press.
Deleuze, G. (1978) *On Spinoza*: http://deleuzelectures.blogspot.co.uk/2007/02/on-spinoza.html (accessed 8 September 2017).
Donnelly, N., Poynton, S. and Weatherburn, D. (2017) *The Effect of Lockout and Last Drinks Laws on Non-domestic Assaults in Sydney: An Update to September 2016*. Crime and Justice Bulletin No. 201. Sydney: NSW Bureau of Crime Statistics and Research.
Dumas, D. (2016) 'Going, going, gone: 10 iconic bar closures and moves in Sydney', *Sydney Morning Herald*, 14 March: www.smh.com.au/nsw/going-going-gone-10-iconic-bar-closures-and-moves-in-sydney-20160314-gnic6h.html (accessed 19 October 2017).
Guattari, F. (1993) 'Space and corporeity', *Columbia Documents of Architecture and Theory*, 2, pp. 139–149.
Gwiazdzinski, L. (2005) *La Nuit, dernière Frontière de la Ville*. La Tour-d'Aigues: Editions de l'Aube.
Harvey, D. (2012) *Rebel Cities*. London: Verso.
Hobbs, D., Hadfield, P., Lister, S. and Winlow, S. (2003) *Bouncers: Violence and Governance in the Night-time Economy*. Oxford: Oxford University Press.
Hollands, R. (1995) *Friday Night, Saturday Night*: http://research.ncl.ac.uk/youthnightlife/HOLLANDS.PDF (accessed 18 October 2011).
Ibrahim, M.F. and Leng, S.K. (2003) 'Shoppers' perceptions of retail developments: Suburban shopping centres and night markets in Singapore', *Journal of Retail and Leisure Property*, 3(2), pp. 176–189.
Jackson, P. (2004) *Inside Clubbing*. Oxford: Berg.
Jayne, M., Gibson, C., Waitt, G. and Valentine, G. (2012) 'Drunken mobilities: Backpackers, alcohol, "doing place"', *Tourist Studies*, 12(3), pp. 211–231.
Jayne, M., Valentine, G. and Holloway, S.L. (2011) *Alcohol, Drinking, Drunkenness: (Dis)orderly Spaces*. Aldershot: Ashgate.
Lefebvre, H. (2004) *Rhythmanalysis: Space, Time and Everyday Life*. Translated by Elden, S. and Moore, G. London: Continuum.

Lin, Y.C., Pearson, T.E. and Cai, L.A. (2011) 'Food as a form of destination identity: A tourism destination brand perspective', *Tourism and Hospitality Research*, 11(1), pp. 30–48.

Malbon, B. (1999) *Clubbing: Dancing, Ecstasy and Vitality*. London: Routledge.

Mbonye, M., Nalukenge, W., Nakamanya, S., Nalusiba, B., King, R., Vandepitte, J. and Seeley, J. (2012) 'Gender inequity in the lives of women involved in sex work in Kampala, Uganda', *Journal of the International AIDS Society*, 15(1), pp. 27–35.

Melbin, M. (1987) *Night as Frontier: Colonizing the World after Dark*. New York: The Free Press

Montgomery, J. (1995) 'Editorial: Urban vitality and the culture of cities', *Planning Practice and Research*, 10(2), pp. 101–110.

Morrissey, J. (2005) 'Cultural geographies of the contact zone: Gaels, Galls and overlapping territories in late medieval Ireland', *Social and Cultural Geography*, 6(4), pp. 551–566.

Peck, J. and Tickell, A. (2002) 'Neoliberalizing space', *Antipode*, 34(3), pp. 380–404.

Radley, A. (2003) 'Flirting', in Coupland, J. and Gwyn, R. (eds) *Discourse, the Body and Identity*. Basingstoke: Palgrave Macmillan, pp. 70–86.

Roberts, M. and Eldridge, A. (2009) *Planning the Night-time City*. London: Routledge.

Schlör, J. (1998) *Nights in the Big City: Paris, Berlin, London 1840–1930*. London: Reaktion.

Schulkind, J., Mbonye, M., Watts, C. and Seeley, J. (2016) 'The social context of gender-based violence, alcohol use and HIV risk among women involved in high-risk sexual behaviour and their intimate partners in Kampala, Uganda', *Culture, Health and Sexuality*, 18(7), pp. 770–784.

Ssewakiryanga, R. and Isabirye, J. (2006) '"From war cacophonies to rhythms of peace": Popular cultural music in post-1986 Uganda', *Current Writing: Text and Reception in Southern Africa*, 18(2), pp. 53–73.

Stepney, M. (2014) 'British women's experiences of drinking alcohol: Dynamics of emotional proximity and distance', *Emotion, Space and Society*, 10, pp. 105–112.

Szmigin, I., Griffin, C., Mistral, W., Bengry-Howell, A., Weale, L. and Hackley, C. (2008) 'Re-framing "binge drinking" as calculated hedonism: Empirical evidence from the UK', *International Journal of Drug Policy*, 19(5), pp. 359–366.

Wasswa-Matovu, J. (2012) 'Youth livelihoods and karaoke work in Kampala's nightlife spaces', in Bourdillon, M. and Sangare, A. (eds) *Negotiating the Livelihoods of Children and Youth in Africa's Urban Spaces*. Oxford: African Books Collective, pp. 151–168.

Wilkinson, S. (2015) 'Alcohol, young people and urban life', *Geography Compass*, 9(3), pp. 115–126.

Winlow, S. and Hall, S. (2009) 'Living for the weekend: Youth identities in northeast England', *Ethnography*, 10(1), pp. 91–113.

Yu, S.-D. (2004) 'Taiwan's night market culture', in Haddon, R., Jodan, D.K., Morris, A.D. and Moskowitz, M.L. (eds) *The Minor Arts of Daily Life: Popular Culture in Taiwan*. Honolulu: University of Hawaii Press, pp. 129–149.

5 Aesthetics of the night-time city

In Steven Spielberg's 2016 version of *The BFG*, an adaptation of the novel by Roald Dahl, the opening scenes show the title character – the Big Friendly Giant – patrolling the streets of London at night, attempting to avoid being seen by people who are out and about. Spielberg's London at night is a space inhabited by many of the people we have encountered so far in this book: drinkers, delivery workers, police officers and a few animals. The BFG hides parallel to lampposts, uses his cloak to cast shadows, and plays with the dappled landscape of light and dark that is produced at night to move largely unnoticed through the city. Although fanciful, the London of *The BFG* is instantly recognizable, drawing as it does on depictions of the night as a quiet time in which a small number of actants navigate a changing landscape of light and dark while the majority of people rest. Despite being, as we have explored, a rather peripheral space within city life, the urban night appears frequently in a variety of different media. From advertising to poetry and film to song, the city at night is often presented as an essential part of urban living, a moment when the city is distinguished from the rural, and large cosmopolitan cities are distinguished from smaller urban environments. At the same time, as in *The BFG*, representations of the nocturnal city often use it as a timespace in which something is different, be it danger, mystery or the uncanny.

How does this happen and what does it tell us about both night and the city? In this chapter I will explore in more depth the representation of the urban night, looking both at how the night appears in culture and how this connects to lived experiences of the city. Crucially, as well as focusing on acts of representation, I am interested in the normative values that these representations encompass and their use in the promotion and encapsulation of urban life. It is for this reason that I use the term 'aesthetics' – to capture the role of how the night appears or is experienced in producing the norms of the night. In other words, this is not just a case of representations producing discourse, but a series of aesthetic tools – lighting, colour, sound, sensation – which make social practices and shape urban planning. I track this across urban lighting spectacles in the nineteenth century, the night-walking of Dickens and contemporary urban explorers, and the use of night in film noir and connected art forms. I then turn to the use of the night in marketing and civic boosterism, in which projections or images of night-time cityscapes are employed to illustrate dynamic and cosmopolitan cities.

I believe there is a direct connection between the ways in which cities at night are represented, the ways in which they are lived, and the ways in which the representations are used. In other words, representations of city nights are 'performative': that is, they gain meaning through their lived enactment and performance. To paraphrase Butler (1988, p. 527), they are real only to the extent that they are performed. As urban inhabitants encounter the night-time city, they produce both the city itself and representations thereof; these representations then help to make the city at night what it is.

From an aesthetic perspective, I am interested in how an understanding of the aesthetics associated with the night-time city can aid an understanding of 'the body and its place within the world' (Hawkins and Staughan, 2016, p. 3). Aesthetics informs us about what is valued in the experience of the night-time city, both in the liminal moments of festivals and in the mundane temporalities of the everyday. In particular, I want to tease out the relationship between, on the one hand, depictions of the night-time city which revel in the 'night as spectacle' – a canvas of lighting and performances that might amuse or enchant but from which the observer is broadly detached – and, on the other, the 'night-time city as spectacular' – a fascinating and alluring environment in which one is invited to lose oneself. In turn, these different forms of aesthetic representation and living reveal how the night-time city can be simultaneously central to urban life and on its borders.

Illuminations and conviviality: developing a night-time aesthetics

Returning to the medieval night that Ekrich describes, we find a time in which the night was represented as a timespace of danger and threat. In art and poetry, the night and darkness were inherently connected to Hell and demons: so Ekrich (2006, p. 14) cites the Stuart playwright John Fletcher, who describes demons as the 'blacke spawne of darknesse'. Darkness also appears repeatedly in Dante's *Inferno*, which opens with the scene-setting lines: 'In the midway of this our mortal life, I found me in a gloomy wood, astray'. Dante uses a literary trope here that continues through nineteenth-century Gothic texts and into contemporary horror writing, films and games, whereby open scenes find a protagonist alone and isolated in the darkness – soon to be tricked, dragged or led into danger and threat. Darkness has also fed directly into the aesthetics of literature. In 1816, Western Europe experienced a dark 'Year without a Summer', caused by the presence of ash in the earth's upper atmosphere following the eruption of Mount Tambora in what was then the Dutch East Indies. The subsequent experience of gloom, darkness and extended night can be seen in the art and literature produced at the time. Most famously, Mary Wollstonecraft-Shelley, John William Polidori, Lord Byron and others spent the summer writing in a villa near Lake Geneva. From this, the darkness-drenched *Frankenstein* by Wollstonecraft-Shelley, *The Vampyre* by Polidori and *Darkness* by Byron all eventually emerged. In the Gothic aesthetics of this work, we see the dark night of 1816's direct influence on culture.

This Gothic aesthetics of night and darkness has clearly continued into the contemporary era. However, despite remaining an important and influential depiction of the night-time city, the long-held tradition of tales and imagery of the night as dangerous and threatening is no longer the only nocturnal story to be told. Koslofsky locates the start of the decline of this tradition in European culture in the seventeenth and eighteenth centuries. Citing English poetry from the time, she argues that this period saw

> the use of the night in political spectacle and the increase in the scope and legitimacy of everyday nocturnal activity. One can refer to the sum of these processes, symbolic and quotidian, as 'nocturnalization', a decisive step in the development of the modern night.
> (Koslofsky, 2007, p. 236)

Such developments were first found in royal and political courts, where wealth and power brought access to developing technologies of artificial illumination. Louis XIV of France, the 'Sun King', is first recorded as adopting this moniker through his appearance in the illuminated 'Ballet de la Nuit' in 1653, which used technological innovations in lighting and staging (Koslofsky, 2007, p. 239). Thereafter, lighting was employed in the night-time festivals of the later seventeenth century (Schivelbusch, 1988), which were precursors of the development of night-time leisure societies in the eighteenth and nineteenth centuries (Schlör, 1998). Through these festivals, new ways of engaging with night and darkness emerged that would spread into public life.

So we see an evolution in the depiction and representation of night occurring alongside the growth in night-time society. With increased engagement with the night, an aesthetic of the spectacular emerged, in which night-time cities were vibrant, colourful places where people could encounter new cosmopolitan experiences. This was an embodied, performative aesthetic in which participants were expected to revel (Hawkins and Staughan, 2016). Schivelbusch (1988) offers numerous contemporary descriptions of nineteenth-century lighting festivals and projections, revealing further the fascination and wonder that such projects generated. For instance:

> In the middle of the night, we emerge into the brightest daylight. Shop and street signs can be recognised clearly from across the street. We can even see the features of people's faces well from quite a distance, and what is especially remarkable, the eye accustoms itself to this intense light immediately.
> (De Cluny, 1880, quoted in Schivelbusch, 1988, p. 118)

Inventors, entrepreneurs and showmen (and it was always men) travelled across Europe and North America selling ever more ambitious lighting projects and structures. We see here an increase in the emphasis on *light* in architecture and urban design, an emphasis that would increase in the modernist architecture of the early twentieth century (Steane, 2011). This approach to lighting reflects a broader modernist sense of progress in which technological improvements are credited with the capacity to enhance urban life. The aesthetics of the urban night that developed around this time, then, brought with it efforts to open up and develop a lived aesthetics of conviviality within the night-time city. In turn, this conviviality was important in shaping the image of the urban night. As Edensor (2015, p. 426) argues, 'rather than remaining within their homes through fear of the dark, people flooded into urban streets in search of amusement, spectacle, commerce and new forms of conviviality'. As previously discussed, a new form of night-time leisure was created, and this was tied to the aesthetics of the urban night as a brightly lit playscape.

These shifting representations of the night-time city can be seen in literature, too. We can see, for example, a shift in the depiction of the state at night and in particular the security services who work for the state. In early modern and earlier times, nights were governed by nightwatchmen. In popular culture, these figures were often considered rather hapless and incompetent, as is seen in the character of Dogberry in Shakespeare's *Much Ado about Nothing*. He and his fellow nightwatchmen are self-important constables, prone to sleeping through the night shift. Their work is successful only by chance, and Dogberry, the captain of the squad, is particularly foolish. The nightwatchmen were the lowest of the state's security forces and as such were considered to be among the weaker of the state's officials; in Shakespeare's time, Dogberry would have been understood as a commentary on this (Spinrad, 1992). Night was a time of danger in which the state's authority largely evaporated, leaving only its most incompetent officials to exert a modicum of control. By contrast, the newly professional police constables of the nineteenth century were highly respected, and their knowledge of the night was considered advantageous. The early police forces sought to respond to a newly lively night through much more proactive and engaged techniques of governance and management (Schlör, 1998). In literature, we also see the emergence of detectives who are comfortable with the night-time city, of which the best-known example is Arthur Conan-Doyle's Sherlock Holmes. He moves freely around the darkened spaces of the city, working with the inhabitants of the night-time urban landscape to understand and thwart its criminal elements. This reflects the state's changed approach to the night-time city, which by this time it was seeking to observe and control rather than prohibit. In literature, we thus find a connection between representations of the night-time city and practices of nocturnal securitization and governance.

This section has told the story of the diversification of representations of the urban night, and the parallel development of new aesthetic practices. Before the Enlightenment, the night was mainly a time of danger and myth. While these discourses have not disappeared, they were joined in the eighteenth century by night spectaculars, festivals, illumination and a modern fascination with expansion (Edensor, 2017). The night shifted from a timespace dominated by depictions of darkness to one which also contained representations of revelry. Bound up with this was a shift in how the night-time city was managed and governed. So we see in this transition that the history of the development of the urban night as 'spectacular' was inherently connected to a wider opening up of the urban night to facilitate greater activity and tighter governance. In other words, this is what we might call an 'ethico-aesthetics' (Guattari, 1995), in which the ways of inhabiting the night properly are encouraged and governed by an aesthetic imagining.

Walking, exploration, graffiti: a counter-aesthetics of the night?

We can find, however, a series of lived aesthetic practices which offer different ways of inhabiting the night-time city. In this section, I want to posit five different expressions of a 'counter-aesthetics' of the night, all of which locate people in the city at night in ways which go beyond or outside of the 'spectacular'. Instead, this form of aesthetics might be described as the urban night as 'spectacle', rendered open to the – usually white, cisgender male – observer to gaze upon in fascination.

Researchers across various disciplines have explored the ways in which writers in particular have drawn on the practice of walking through the city at night as a means of producing urban literature. Beaumont's previously introduced work on night-walkers describes the outsiders who have been pushed towards night-walking throughout (European) history. He differentiates between those who are forced into the night – 'the sad, the mad, the bad' (Beaumont, 2015, p. 3) – and those whom he terms 'noctambulants' – tourists who have 'made excursions into the nocturnal streets because … [they have] chosen to do so, not because … [they have] been exiled there by circumstance' (p. 138). In the former camp, Beaumont identifies individuals for whom the night has been a space of both refuge and (rather dangerous) opportunity. In the latter, we see those who have sought out the night as an aesthetic landscape for immersion and inspiration. Beaumont cites authors, poets and playwrights who have all used night-time walking as a source of inspiration. If Coleridge and Wordsworth introduced a method of landscape-centred artistic production in the countryside, then Beaumont identifies a number of writers, most famously Dickens, who did the same in the city at night.

We have already discussed Dickens as one of our 'nocturnals' in Chapter 2, but here we can focus on the aesthetic values he found in the night-time city. He reflected on these in a series of essays published in 1860 as *The Uncommercial Traveller*, and more recently collated in the single-volume *Nightwalks*, named after one of the essays (Dickens, 2010). Here, Dickens describes a generally quiet night-time city, with pockets of activity: evening entertainment venues which gradually reduce in number as the night goes on; train stations at which deliveries from the countryside arrive; church bells ringing to mark the time; and the restocking of markets. Company, in the form of people or through encounters with lit buildings, is usually welcome: Dickens describes one such encounter as 'warming', but also, 'like most of the company to be had in this world [the city at night], it lasted only a very short time' (p. 10).

While Dickens is certainly the best known, Beaumont shows that he was just one of many nineteenth-century writers to use night-time walks as inspiration for understanding the city in more detail. Across this area of literature, the urban aesthetic developed a clear love for the (romanticized) isolation and darkness of the city at night, accessed selectively from a position of privilege. As with the Romantic poets who sought the isolation of the countryside for their inspiration, this approach takes the uncanny urban – the night – flickering in and out of the social world, as its literary prompt, but does not necessarily engage fully with the lives of those who are forced into these spaces.

While night-walking has been a significant practice in the production of urban literature, the figure of the night-walker also appears with some frequency in cinema, particularly in the film noir genre. Indeed, we can see this genre as a direct descendant of the Dickensian night. In a classic text, Place and Peterson (1974, p. 30) identify film noir's characteristic traits of 'claustrophobia, paranoia, despair and nihilism'. These are typically played out in darkened cities at night, to such an extent that film noir has become one of the main ways of imagining the urban landscape on screen (Clarke, 2005; Naremore, 2008). There are numerous examples of this style, but a particularly interesting one is the 2014 film *Nightcrawler*. Many classic film noirs take place in California, reflecting the region's early development of an extensive night-time lighting network, and *Nightcrawler* maintains this tradition by setting the action in Los Angeles. The story follows Lou Bloom, a petty criminal who stumbles across the world of 'stringers' – freelance photojournalists who listen to emergency service broadcasts in order to reach crime scenes before the police – after witnessing a night-time road-traffic accident. Bloom becomes a stringer himself and ventures ever further into the night-time city until he reaches the point of deliberately concocting or encouraging criminal activity that he can film with the intention of selling the footage to the local television news networks. So, as in many classic film noir movies, the lead character – typically a white, male *flâneur* who struggles on the periphery of daytime society – identifies an opportunity which draws him towards the night's seedier elements. As we follow Bloom through the city's streets, the night is laid out in front of us as a spectacle, rendered open to our voyeuristic fascination. Night here shares many characteristics with the frontier – lawless (or at least lacking legal control), lonely and sparsely populated as well as predominantly male.

The depictions of film noir can be seen in other art forms, too. The night-time paintings of Edward Hopper, most famously *Nighthawks*, certainly reflect a film noir sensibility. In this work, three people – a male and female couple and one man on his own – sit around a counter in a diner, with the city outside the window dark and empty: 'the scene is lit by the harsh light from the diner that spills out onto a quiet green street; dreary red buildings fill the background' (Harris, 2006, p. 715). The protagonists are all hunched, and both they and the city seem quiet and isolated, as in film noir. Painted in 1942, the image depicts a lonely city in which the actants are lit by artificial light, which makes them visible to our surveillance. Again, there is a clear connection between this work and Dickens's walks, with both the artist and the writer motivated to tell the stories of the isolated, the lonely or the outcast. Thus, from the mid-nineteenth century to the twentieth century, we see the development of an alternative nocturnal aesthetic, centred on a romanticization of lonely wanderings in the dark. This vision of the night-time city, refracted through Victorian-era writing and (re)emerging in film noir and the visual arts, typically casts the reader or viewer as an explorer – someone who discovers an environment that is at once both familiar and uncanny: their normal city transformed into a space inhabited by 'the sad, the mad, and the bad'. The film noir aesthetic rarely places us in the position of these night-walkers themselves; rather, we are the observing noctambulants. This is not the night-time city as *spectacular*. Instead, it revels in a fascination with the dangerous, the dark and the weird, so we might describe it as the night-time city as *spectacle* – laid open for us, the observers, to explore. It becomes a strange, unusual and inherently *different* form of city for us to encounter.

The late twentieth and early twenty-first centuries have produced a new imagining of this form of night-time aesthetics, in which there is a greater physical engagement with the city. 'Urban exploration' has been of increasing interest to academics and artists alike, in part due to the ways in which it merges the night-time city as spectacle with the night-time city as spectacular. Urban exploration encompasses a range of practices related to engaging with the built environment of a city by gaining access to abandoned buildings, tunnels, construction sites and infrastructure in marginal or restricted areas. In a well-known incident, geographer Bradley Garrett made the front pages of British newspapers when he climbed the Shard skyscraper in London with urban explorers while researching his Ph.D. (Garrett, 2013). To illustrate story, most of the papers used an image of Garrett atop the Shard with night-time London stretched out in front of him. Delving further into his and other researchers' work, we find that urban exploration is often illustrated with nocturnal images (Pinder, 2005; Paiva, 2013). For example, Garrett's book features numerous photographs of urban landscapes illuminated by artificial light, some of which include the geographer himself, looking down over the city (see Garrett, 2013, pp. 75–94).

Urban exploration has been pushed into the night for several obvious reasons, such as the reduced risk of surveillance, the fact that many urban explorers work or study during the day, and the advantages afforded by the cover of darkness. However, as a practice that has come to prominence alongside digital photography, which has allowed the explorers to share images of their activities, the rise of urban exploration has combined both the aesthetics of the spectacle and the aesthetics of the spectacular. The shared photographs are certainly spectacular, often focusing on the brightly lit urban carpet that spreads beneath the explorer. However, by placing the explorer front and centre, looking out over the city, they also contain elements of the city at night as spectacle. Urban explorers are thus both individuals who are forced into a dark and dangerous night-time city to undertake their work *and* the producers of images of the spectacular night-time city. This practice therefore merges the spectacle and the spectacular to produce engaging imagery of the city that is nonetheless still rendered as object for the viewer.

Cities at night and place marketing

So far, this chapter has focused on representations of the city and their connection to the lived landscapes of night. I have argued that, as well as the darkness and intrigue of the classic 'Gothic' night-time city, depictions of night have treated it as both spectacular – as a space of embodied possibilities – and as spectacle – as a space of fascination. These representations reinforce the sense of night as a border to urban life: something that is both within the city and around its edge. In the night as spectacular and the night as spectacle, the uncanny night-time city is somewhat outside of normal forms of urban life. However, as places accessed daily by urban inhabitants – whether as revellers in the night as spectacular or noctambulants in the night as spectacle – these aesthetics also position the night as part of urban life.

In this section I want to develop this second aspect of urban nocturnal aesthetics by looking at the role that night plays in globalized place-marketing strategies. Alongside the rise in lighting festivals that was discussed in Chapter 3 (see also Edensor, 2017), this reveals a night which is central to globalized urban life.

The concept of place marketing emerged in the 1990s as part of the 'entrepreneurialization' of cities. This was a decade when being an 'entrepreneurial city' became 'a central element in many cities' self-imaging' (Jessop, 1998, p. 77). Entrepreneurialization entails the city promoting the capacities of its 'economic spaces in the face of intensified competition in the global economy' (p. 80) and marketing itself as attractive to an imagined global population of highly mobile, Westernized elites. Florida (2002) describes a sub-set of these elites as the 'creative class': individuals employed in precarious or short-term but relatively high-income professions, typically but not exclusively associated with the arts, leisure and heritage. From here, he develops a strategy for good urban growth that has proved highly influential, but which is also problematic (Peck, 2005). Crucially, Florida presumes that attracting the creative class is invariably a 'good thing' as he explores the group's economic productivity and cultural capital while overlooking their gentrifying impact on urban life. However, he does identify the cultural and economic values that are associated with this group of creatives. Furthermore, there is strong evidence that their cultural values, at least, are shared by a wider global middle class, including those who work outside the creative industries (María Luisa Méndez, 2008). As such, Florida's work usefully identifies many of the cultural preferences of the elusive 'footloose' middle class, a group which the world's entrepreneurial cities are keen to attract.

An attractive, open and vibrant nightlife is one of the key values that is associated with this global middle class (Binnie and Skeggs, 2004). This has led to cities across the world producing spaces that facilitate such a nightlife, often seeing an equivalence between nightlife and 'cosmopolitanism'. Hae (2011, p. 3461) describes this as 'the nightlife fix', in which an active night-time alcohol and leisure industry is imagined to have self-perpetuating benefits: nightlife attracts a young and wealthy crowd, who chose to locate their entrepreneurial businesses in a city, which in turn drives the production of a more innovative and vibrant nightlife, and so forth, in a virtuous circle. In other words, a healthy nightlife is seen as a key means of attracting young professionals and creatives to a city. The result is that the urban night often features prominently in place marketing which targets such groups. Furthermore, the city at night is used in the branding and imagery surrounding places in a wider context. In the brochures and leaflets that promote cities, images of the night are used to illustrate many of the positive features of urban life: vibrancy, connectivity, youthfulness and a global or Western orientation. The aesthetics of the night as spectacular can be seen in the context of a wider planetary, cosmopolitan, urban aesthetics, in which a hegemonic view of the 'global city' influences the planning and management of cities.

A leaflet entitled 'At Night' – produced for a particular British town (name briefly redacted) – provides an insight into the ways in which marketing agencies use nightlife to attract people to urban areas:

— is a vibrant university town … It's at the heart of everything with an exciting programme of live music, comedy, theatre and exhibitions and an emerging craft ale and foodie scene … Lose yourself in one of —'s large clubs, sip cocktails in a stylish bar or enjoy real ale in one of —'s brilliant pubs … With a growing choice of eateries on Linthorpe Road, there really is a buzz in the air.

The leaflet is illustrated with images of the town's landmarks at night or as the sun goes down, interspersed with what appear to be stock images of bars and clubs. But what is this 'vibrant' town that is 'at the heart of everything'? The answer is Middlesbrough – a peripheral industrial town in England's northeast which saw a 2 per cent population decline between 2001 and 2011, and in which half of the electoral wards featured in the top 10 per cent of the British government's 'Index of Multiple Deprivation' in 2015. Indeed, one of Middlesbrough's wards was second and another was tenth on that depressing list (Tees Valley Unlimited, 2015). In other words, its self-proclaimed status as a 'cosmopolitan', 'vibrant' town is rather questionable. But for our purposes, what is most striking here is that Middlesbrough chooses to market itself on the basis of its nightlife. Even though it struggles to attract both investment and visitors, it presents images and a discourse that draw heavily on conceptualizations of the urban night that the town cannot hope to meet. Rather than focusing on what Middlesbrough *can* offer, the place marketing is driven by an aesthetic understanding of what night-time cities *should* offer.

In placing the night-time city in this position, we see elements of both spectacular and spectacle gaining other discursive meanings. The urban night as spectacle is offered through the 'sophisticated' and 'stylish' aspects of city life, which appeal to a mainstream tourist audience for whom the ability to sit and watch an attractive, exciting night-time cityscape is as important as experiencing it. However, if place marketing is to tempt the all-important creative class, it must also offer the night as spectacular by representing nightlife as 'edgy'. In Amsterdam, for example, the nightlife industry lobbied for the creation of a 'Night Mayor' – 'a rebel in a suit' who would represent the edgy elements of nightlife in the political mainstream. Similarly, in November 2016, the DJ Amy Lamé became London's first 'Night Czar'. Her official Twitter account describes her goal as 'making #London the most diverse & dynamic 24hr city in the world. Nightlife IS my life.' While Lamé's appointment reveals a willingness to speak to the edgy night, it should be noted that her role is supported by the Office of the Mayor of London, which places it firmly within the city's official place-branding strategy. In Singapore, the night-time economy has been developed through lavish venues in an 'effort to reimage Singapore as a more vibrant and fun-loving city' (Tan, 2012, p. 723). Here, strip clubs, a relaxation of alcohol legislation and other measures have been used in a bid to attract a globally mobile audience. All of these examples place *experiencing* the urban night – the night as spectacular – at the centre of planetary urbanization. However, they simultaneously emphasize the spectacle of the urban night with the aim of attracting those who might want to observe rather than participate.

But it is not just in the production of nightlife that the central role of the night in a cool urban image can be seen. Architectural visualizations can be a useful way of identifying what is viewed as essential to urban life (Rose *et al.*, 2014), and one interesting aspect of these is the prevalence of pictures of the city at night within imagery of new urban developments (see, for example, the sunset image which Degen *et al.* (2015) use as a key case study). At the time of writing, websites publicizing future developments in a variety of global cities contain numerous images of these sites at night: for example, Baltimore Tower, London (Baltimore Tower, 2017); Guozen Securities Tower, Shenzhen (Fuksas, 2017); and the Pacifica, Auckland (Pacifica, 2017). In all of these cases, night-time imagery advertises the spaces as lively and vibrant, which places the night at the heart of future developments.

So, as cities have had to compete on a global scale, the circulation of the urban night as both spectacle and spectacular has contributed to a homogenization of urban strategies across inter-city networks. Such networks encourage cities to replicate strategies, often with the goal of emulating the perceived successes of global cities. These strategies place the night at the heart of what it means to be urban, even though much of the value of the aesthetics of the night-time city rests on the ways in which it falls outside of the urban norm. Therefore, place marketing exploits the night's capacity to convey a distinctly urban form of living at a time when the city loses most of the characteristics that define urban life. As Sharpe (2009, p. 2) puts it, 'a shimmering skyline and a blaze of electricity signified human life at its richest, most promising and most seductive: "bright lights, big city"'. The brightly lit night-time city has become a key image of successful urban life.

The aesthetics of the night and what it means to be urban

In this chapter, I have claimed that we can identify a distinctive aesthetics of the night in the use of imagery, discursive understandings and lived engagement with urban night-time space. This cuts across both representation and inhabitation of the city (Hawkins and Staughan, 2016), revealing the normative meanings associated with understandings and performances of space. In doing this, we have covered both the aesthetic use of night-time city imagery and the aesthetics of embodied engagement with the urban landscape. These aesthetic approaches can be described under two 'visions' of the night-time city, notwithstanding that actual urban representations are usually multifaceted and contain elements of both:

1 A vision of the night as spectacular – in which the city is inhabited and experienced as wondrous.
2 A vision of the night as spectacle – in which the city is explored and observed by the *flâneur*, often with some of its marginal elements visible.

These different practices and representations combine to produce what we might call the contrasting aesthetics of the frontier. In other words, this aesthetic vision of the night reflects more broadly an aesthetic vision that we see repeated in other visions of the frontier. Imagine, for example, the classic scene in a Western film when the male, white hero strides into town: the space is both dangerous and alluring. If frontier spaces have operated on the border of the rules of the metropole, nocturnal aesthetics reveal the night as a space on the border of the rules of the day. The practices of place marketing and urban exploration reveal the tension in which this dual aesthetic vision sits. On the one hand, the night as spectacular is there to be lived – a vision to admire, certainly, but one which is admired through a deep, embodied experience of space (Morris, 2011). This vision of night contains bright lights set against a backdrop of darkness, which gives them attractive vibrancy. It has elegant bars and a diverse and youthful populace, but also edgy places that offer the dangerous and risky forms of entertainment described in Chapter 4. On the other hand, in the night as spectacle exploration and experience are practised at a distance. The people and places that the *flâneur* encounters are curiosities, oddballs who might reveal the interesting or inspiring aspects of a city, but the visitor observes them without fully engaging with them. In practice, the lived night-time city involves a nuanced intermingling of these elements of a nocturnal-frontier aesthetics, but identifying them is nonetheless useful.

This reveals three central aspects of the night and the city. First, it establishes some of the ways in which the night-time frontier will fragment and splinter. Following postcolonial theory, which has defined frontiers as contact zones, we can argue that these differing aesthetics operate in the same way: as contact zones of different forms of viewing and engaging with the city. The aesthetics of the urban night, which pivot between spectacular and spectacle, inherently pivot between the centre and the margins, the light and the dark, the day and the night. This encourages the fragmentation of the frontier. Second, these aesthetics reiterate the night's centrality to common understandings of what it means to be urban. The prevalence of images of a lively and active night in promotional material is testimony to that. So, the urban night is a particularly urban moment, and the aesthetics associated with that reveal something about how we *should* live in cities. Third, the nocturnal urban aesthetics reveal the persistence of something different within the city. Some parts of every city are viewable, with lights and displays; but other parts are quiet backwaters that are difficult or dangerous to explore. Garrett and other urban explorers, place marketers who promote nightlife districts, and artists, writers and poets who still walk through cities at night for inspiration (see, for example, Self, 2014) all provide evidence of a night-time city which offers something different from the day in ways that are both imagined and real. In other words, for this distinctive night-time aesthetics to exist, the night-time city must retain its independence and distinction from the daytime city.

The aesthetics of the urban night, as explored here, is the final element of the 'public' night that I cover in detail in this book. I hope it reveals the global reach of images and understandings of night-time cities. Returning to the understanding of the planet that we developed out of Guattari's ecosophy, however, the global should be understood only as one manifestation of the planetary. An ecosophical urbanism also pays attention to the direct interactions that we have with the planet as our home (the etymological origins of the prefix 'eco-'). To that end, the next chapter explores the domestic night as a different form of nocturnal urban space.

At this point, then, it seems useful to provide a summary of the last three chapters on the public night. I have argued that the urban night is uniquely distinct in its infrastructure, liveliness and representation. This distinction has increased as capital has expanded into the night, making it a 'frontier'. This frontier is moving ever further from the biogeoastronomical night that has characterized the hours of darkness on earth for centuries. However, I have also explored the ways in which the frontier is fragmenting, or perhaps is necessarily fragment*ed*. In other words, the night that is mainstream is also peripheral; the night that is spectacular is also spectacle; the night that is economically productive is also empty. While various forces therefore homogenize the night in cities, and make it more planetary in the sense of 'global', the same forces make it less planetary in the sense of connections to the earth, and do not fully penetrate the city. In these terms, we can continue to understand night as the expansionary point of capital in the sense of both frontier and border.

References

Baltimore Tower (2017) Website: www.baltimorewharf.com/index.html (accessed 20 October 2017).
Beaumont, M. (2015) *Night Walking*. London: Verso.
Binnie, J. and Skeggs, B. (2004) 'Cosmopolitan knowledge and the production and consumption of sexualized space: Manchester's gay village', *Sociological Review*, 52(1), pp. 39–61.
Butler, J. (1988) 'Performative acts and gender constitution: An essay in phenomenology and feminist theory', *Theatre Journal*, 40(4), pp. 519–531.
Clarke, D. (2005) *The Cinematic City*. London: Routledge.
Degen, M., Melhuish, C. and Rose, G. (2015) 'Producing place atmospheres digitally: Architecture, digital visualisation practices and the experience economy', *Journal of Consumer Culture*, 17(1), pp. 3–24.
Dickens, C. (2010) *Night Walks*. London: Penguin.
Edensor, T. (2015) 'The gloomy city: Rethinking the relationship between light and dark', *Urban Studies*, 52(3), pp. 422–438.
Edensor, T. (2017) *From Light to Dark*. Minneappolis: University of Minnesota Press.
Ekrich, A.R. (2006) *At Day's Close*. London: Pheonix.
Florida, R.L. (2002) *The Rise of the Creative Class*. New York: Basic Books.
Fuksas (2017) 'Guosen Securities Tower': www.fuksas.it/en/Projects/Guosen-Securities-Tower-Shenzhen (accessed 20 October 2017).
Garrett, B.L. (2013) *Explore Everything: Place-hacking the City*. London: Verso.
Guattari, F. (1995) *Chaosmosis: An Ethico-aesthetic Paradigm*. Translated by Bains, P. and Pefanis, J. Sydney: Powet.
Hae, L. (2011) 'Dilemmas of the nightlife fix: Post-industrialisation and the gentrification of nightlife in New York City', *Urban Studies*, 48(16), pp. 3449–3465.
Harris, J.C. (2006) 'Nighthawks', *Archives of General Psychiatry*, 63(7), pp. 715–716.
Hawkins, H. and Straughan, E. (2016) 'For geographical aesthetics', in Hawkins, H. and Straughan, E. (eds) *Geographical Aesthetics: Imagining Space, Staging Encounters*. London: Routledge, pp. 1–18.
Jessop, B. (1998) 'The narrative of enterprise and the enterprise of narrative: Place marketing and the entrepreneurial city', in Hubbard, P. and Hall, T. (eds) *The Entrepreneurial City*. Chichester: Wiley, pp. 77–99.
Koslofsky, C. (2007) 'Princes of darkness: The night at court, 1650–1750', *Journal of Modern History*, 79(2), pp. 235–273.
María Luisa Méndez, L. (2008) 'Middle class identities in a neoliberal age: Tensions between contested authenticities', *Sociological Review*, 56(2), pp. 220–237.
Morris, N.J. (2011) 'Night walking: Darkness and sensory perception in a night-time landscape installation', *Cultural Geographies*, 18(3), pp. 315–342.
Naremore, J. (2008) *More than Night: Film Noir in Its Contexts*. London: University of California Press.
Pacifica (2017) Website: http://thepacifica.co.nz (accessed 20 October 2017).

Paiva, T. (2013) *Night Vision: The Art of Urban Exploration*. San Francisco: Chronicle Books.

Peck, J. (2005) 'Struggling with the creative class', *International Journal of Urban and Regional Research*, 29(4), pp. 740–770.

Pinder, D. (2005) 'Arts of urban exploration', *Cultural Geographies*, 12(4), pp. 383–411.

Place, J.A. and Peterson, L.S. (1974) 'Some visual motifs of film noir', *Film Comment*, 10(1), pp. 30–35.

Rose, G., Degen, M. and Melhuish, C. (2014) 'Networks, terfaces, and computer-generated images: Learning from digital vsualisations of urban redevelopment projects', *Environment and Planning D: Society and Space*, 32(3), pp. 386–403.

Schivelbusch, W. (1988) *Disenchanted Night*. Oxford: Berg.

Schlör, J. (1998) *Nights in the Big City: Paris, Berlin, London 1840–1930*. London: Reaktion.

Self, W. (2014) 'If you want to see London with completely new eyes, take a night-hike out of town', *New Statesman*, 18 September: www.newstatesman.com/culture/2014/09/will-self-if-you-want-see-london-completely-new-eyes-take-night-hike-out-town (accessed 20 October 2017).

Sharpe, W. (2009) *New York Nocturne*. Princeton, NJ: Princeton University Press.

Spinrad, P.S. (1992) 'Dogberry hero: Shakespeare's comic constables in their communal context', *Studies in Philology*, 89(2), pp. 161–178.

Steane, M.A. (2011) *The Architecture of Light: Recent Approaches to Designing with Natural Light*. London: Routledge.

Tan, Q.H. (2012) 'Flirtatious geographies: Clubs as spaces for the performance of affective heterosexualities', *Gender, Place and Culture*, 20(6), pp. 718–736.

Tees Valley Unlimited (2015) *Index of Multiple Deprivation 2015 – Borough Level Results*: https://teesvalley-ca.gov.uk/wp-content/uploads/2016/03/4.-imd_borough_report_2015.pdf (accessed 5 May 2017).

6 The domestic night

When I was a child, I had an elaborate routine for getting from my bedroom to the bathroom at night without ever finding myself in total darkness. This routine was based on the entirely rational premise that, as long as I was in some sort of light, then whatever evil force or monster was lurking in the shadows – and, to be honest, I did not have any particular conception of what that danger might be – would be unable to get me. To reach the bathroom, I had to walk out of my room and down a corridor before turning left and continuing straight on, with the bathroom in front of me at the next corner. What concerned me most was the dark corridor that was to my right as I entered the bathroom. I was pretty convinced that the monsters would be residing down there. Thankfully, though, I had my routine. First, I would switch on my bedside lamp. After casting a careful eye around the room, I would get up and put on the main light. Opening the door to the landing, I would gradually speed up as I moved away from my room. At the point where I turned to the left, the door to the bathroom was about ten paces in front of me. Here, often with one look back to my room, as if to absorb the safety of the light, I would bolt, never daring to look to my right, down the dark corridor, as I tumbled into the bathroom and switched on the light before quickly locking the door. When leaving, the same routine was completed in reverse. I would open the bathroom door and take a moment to compose myself. Then, after switching off the light, I would dash to the corner of the corridor to give myself a view of the safety of my lit room. I would slow down a little as I neared the bedroom door, but would not stop running until I was inside with the door shut behind me. Mission accomplished, I would turn off the main light and then my bedside lamp once I was safely ensconced in my bed.

I followed this routine for the best part of ten years, up to some point in my mid-teens. The home at night was, for me, a space transformed. Gone were the protections and safety that I felt during the day; instead, it was a space of (irrational) fear. This fear is shared by many young children. And, in different, more insidious forms, the home at night can be a space of danger and fear for children and adults alike. For instance, as this chapter will explore, it can be a space of domestic violence or social isolation. However, it can also be a space in which people learn to make and enhance friendships and intimate relationships. It can be a space of comfort and relaxation.

So, how does the domestic night attain these contrasting characteristics? And what is it about the night that alters the home? In this chapter, I will explore the ways in which our experiences of the dark can transform the familiar spaces of home, using phenomenological arguments that help to explain the production of subjectivity in these spaces. I will also argue that the socio-cultural context of the urban night pervades the home, with a lack of support networks and services enhancing isolation. How we experience the home at night is connected to our control over these factors, so, broadly speaking, those who have less control over it as a timespace tend to develop negative and fearful experiences of it. In the case of domestic violence, those who exercise this control often use it to enhance the damage that violence can cause.

So far in this book, I have focused on those public spaces that have been identified in the literature as sites that shape the city at night. From the maintenance workers who are out on the roads, to the drinkers who fill late night bars, pubs and casinos around the world, to the iconic images of city landmarks lit by multicoloured artificial lighting, we have seen the multifarious ways in which the public night-time city has developed and expanded. To dive further into the night as a border to planetary urban life, I now want to move into a timespace that researchers visit less often – the home at night. This will continue our theme of exploring how subjectivity is produced in the urban night. In particular, the home at night might represent a timespace in which those features that apparently define the city – connectivity, encounters with difference, a multiplicity of available services – are absent or at least reduced in many places. Indeed, in this chapter, I argue that the defining features of the night at home are predicated on the absence of many of the ways in which urban life operates. As such, the home at night might fall outside of the 'planetary urbanism' that this book has explored up to this point.

Shaping the domestic night: from comfort to control

Homes vary globally, and anthropological, geographical or sociological studies of different 'domestic nights' perhaps unsurprisingly reveal a wide range of different practices (Bille and Sørensen, 2007; Kumar, 2015a; Bille, 2017). In this section, I want to explore the role of control in shaping the domestic night by telling the story of how the home has become more connected to the state. The focus will be on anglophone Western countries, and the UK in particular. From a more global perspective, similar trends are identifiable in many countries (as the studies cited above illustrate), but there are also variations which mean that the claims made here are not universally applicable.

If we were able to visit a typical British home in the early nineteenth century, we would find it to be largely disconnected, or 'off-grid', as we would say now. At night, the home was almost completely cut off from the world – only a little moonlight, breeze and the odd flying insect could get in, and perhaps only water from a drain would trickle out. Though forms of plumbing have been found in homes across the world for several thousand years, mass domestic plumbing did not begin until the Industrial Revolution and associated urbanization. This plumbing of most homes was the first step in the integration of the home into an urban system, and it had the effect of tying people into the mass provision of facilities. This connection of homes quickly became as much about the incorporation of people (and nature) into capitalist socio-economic systems as it was about the provision of services (Swyngedouw, 1997). Gas and then electric lighting were much later technologies: the first homes were lit by gas in the 1850s and, as outlined in Chapter 2, domestic electric lighting first appeared in 1879 and spread over the next two decades. Since these developments, the home has become further integrated into a series of public infrastructures: telephone, internet, gas-powered central heating and so on. Such infrastructures have been central to providing the comfort and services that we now associate with the domestic sphere. Thus, in a little under 150 years, the domestic night has shifted – almost imperceptibly but importantly and in many ways – from independent of the state and other people to almost entirely dependent upon a series of flows into and out of the home.

This process was not accidental. At key moments, service providers sought to create and capture markets by integrating homes into their grids: Maille-Petty (2010) argues that a variety of actors, such as electricity providers and lighting manufacturers, tried to shape tastes and trends to encourage the quick take-up of electricity in the home in the middle of the twentieth century, in the same way as technology companies seek to drive innovation today. By generating a series of new aesthetics and ideals for the domestic night, these actors generated a new market for lighting products, among other forms of domestic comfort. The ideal of the well-lit home at night was thus encouraged as part of the further subjugation of the home to the market, of the private to the public. Such ideals are not culturally constant: for example, Bille (2015) outlines the Danish domestic sensibility of 'hygge', a series of practices oriented around the production of a comfortable, cosy and welcoming home. There are several elements to it, but the modulation of lighting is central to the production of a hygge environment, with candlelight and shadows used to produce the desired ambience. Even in such contexts, however, there has been marketization: Danish shops now offer a wide variety of products aimed specifically at the hygge market, and the concept has become part of the broader marketing of 'Scandi-cool' in other countries.

So, the home has become increasingly connected to a series of public infrastructures. However, another form of increased connectivity is also shaping the domestic night. Sleep, for example, has recently become subject to efficiency measures that are facilitating the spread of neoliberal demands on subjectivity. One of the main selling points of 'wearable tech', such as smart watches, is that it promotes better sleep. Numerous apps allow users to measure, record and supposedly improve their sleep patterns. A 2016 article on the technology website CNET titled 'How effective is your sleep tracker?' starts with the confession: 'I'm not a good sleeper ... I aim for seven hours a night and average around six and a half, which my body seems to think is normal.' The rival technologies offer a range of measurement and recording options, but all generate a 'sleep score', which the author of the article argues is 'amazing if you're looking to gamify yet another aspect of your bodily functions'(Healey, 2016). On the other hand, this could be seen as an actualization of Berardi's dystopic description of contemporary capitalism on the individual: 'the soul, once wandering and unpredictable, must now follow functional paths in order to become compatible with the system of operative exchanges structuring the productive ensemble' (Berardi, 2009, p. 192). Even sleep – a state during which mind and body have traditionally been at rest and outside of the practices of everyday life (Harrison, 2009) – has been opened up to societal control and management.

As with sleep, the increased networking of the home has opened up the domestic sphere to the neoliberal forces that are associated with day. Numerous writers have suggested that precarity and neoliberal subjectivity are pushing us towards absorbing the demands of capitalism into our bodily lives (Virno, 2004; Hardt and Negri, 2006; Berlant, 2011; Butler, 2015). As the domestic night becomes more open to these forces, so the affects of incessancy may seep ever deeper into the home.

So, the home at night has become increasingly connected, which has made it a more comfortable place, but also more open to control and manipulation. We could understand this as part of the same colonialist expansion of the day that has occurred in the public night.

Home, subjectivity and night

All of this means that the home has become closer to the state over the last 150 years. This is important as the home is key to subjectivity, and the night is a time when this relationship is reinforced. As the increased connectivity suggests, the home is a social space, albeit one which falls largely outside of the public sphere that is often associated with sociality. In this section, I want to explore the nexus between the home and subjectivity. Earlier, I argued that cities are subjectivity machines, and here I want to explore the ways in which individual subjectivity also emerges out of domestic relations. I suggest that this process of producing the self, producing subjectivity, is anchored in the night and, furthermore, that the ways in which subjectivity is produced at night in the home perhaps run counter to the forces at play in most of urban life.

Blunt (2005, p. 506) offers a succinct description of the home: 'a material and an affective space, shaped by everyday practices, lived experiences, social relations, memories and emotions'. There are two key constituents here – the material and affective – as well as several key producers. Blunt argues that the home must be physically somewhere (a material space), and that this somewhere typically contains a set of objects that are often 'belongings': items over which a person or a family claim particular ownership. As Tolia-Kelly (2004) explores, these objects may have a series of personal or socially symbolic meanings, such that their presence is important even in relatively precarious homes. Buse and Twigg (2014) provide an interesting example of this in their study of the significance of handbags for many elderly women in care homes. As a care home resident is stripped of her privacy in many ways, her bag becomes the reduced materiality of 'home' – a private space for the few remaining objects over which the owner still has domestic control.

The home also resonates with both personal and social meaning – it is an affective space. Paradoxically, perhaps, this affectivity can be independent of the home's materialities; visits to former residences, neighbourhoods or home towns that may have been physically transformed generally evoke a series of emotional connections. Notably, the affectivity of home can also operate independently of temporal and spatial scale: a sense of home may emerge either quickly or slowly; and it may be linked to something as small as a handbag or something as large as a region, a country or even the earth (Sagan, 1994).

Despite the multiplicity of scales and features that make up the concept of home, some popular understandings and discourses have developed. One is the 'home as hearth' or the 'home as castle'. Here, the home forms a protective shell, isolated from the threatening dangers of the public sphere. Within this protective bubble, the home fulfils some of our basic psychological needs for emotional and physical safety. This idea resonates across a range of common understandings, and it is regularly invoked in political discourse as a reason for protecting and fortifying various scales of home. Therefore, it is perhaps unsurprising that initial academic interest in this area centred on attempts to problematize and debunk myths of the home as a space of strength, a safe haven for identity, in the face of significant evidence of more problematic forms of domestic living (Domosh, 1998). In contrast to the home as hearth, a less popular, but now well-established, discourse of the home as a space of isolation and entrapment has emerged. In this conceptualization, the isolation and disconnection from society turn the home into a prison. Homes can be spaces of fear or sorrow: a well-known, and growing, social problem is the isolation of elderly people in their own homes, as social lives become increasingly fragmented. This understanding of home resonates on a larger scale, too: people have started to lament the isolation of rural or small-town life, and of deprived neighbourhoods and communities. Pink et al. (2015) have structured these two descriptions of home – as 'hearth' and 'trap' – as a continuum rather than a binary, but the reality is even more complex: both sets of relations often coexist, such that a caring and protective home space simultaneously traps and isolates the person who lives there. This is perhaps best illustrated by research into the experiences of young mothers in the home. Head (2005) suggests that, while they appreciate the opportunity to bond with their babies, some mothers find the home to be an isolated space when they are away from work and engaged predominantly in childcare, with reduced mobility and few opportunities for social contact. A key task of research into the home is to explore the ways in which this sense of isolation starts to emerge, with the ultimate aim of devising ways to combat it.

Researchers have rarely focused on the specific role of the night when exploring the relationship between subjectivity and home. Nevertheless, a number of studies provide some useful information. Many of our uses of the home occur predominantly at night: in the UK, *The Time Use Survey* has found that the most common domestic activities fall into two categories, both of which could be described as broadly 'nocturnal' (Lader et al., 2006). The first category includes 'care of the self' activities: sleep, rest, restful leisure activities (e.g. watching TV), washing, dressing and other activities that help relax and reinvigorate the body. The second category comprises 'home-making' activities, such as cleaning and cooking. While none of these activities is exclusively 'nocturnal', we might reasonably assert that they all tend towards the evening and night in most homes. Furthermore, there is some evidence that they are aligned with the evening and night-time domestic energy-use peaks (Powells et al., 2014), which reinforces the commonsense knowledge that we have about activities such as sleep and television watching. Both categories of activities are key to maintaining the home as a material and affective space. Care of the self activities are dependent upon the affectivity of the home, and specifically the privacy that this creates: we typically want to do them away from the view of others. Home-making activities obviously have a key role in making a house 'homely', but they are also necessary to maintain the materiality of the home. While I do not wish to make assumptions about what 'everybody' does, I think the data reveals that many people spend a large proportion of the evening and night entirely or partially in the home, engaged in activities of self and social reproduction.

So, the night is a key part of our time at home. Our homes at night bind us with those who are alongside us, and separate us from people who are not there. This is not to say that the domestic night is completely cut off and isolated. As outlined previously, there has been some expansion of the day and the public into the night-time home. Globally, there is significant variation in the connectivity of domestic night. Contrast Møhl's study of villages in the Berry region of central France with Kumar's research in Bihar, northern India. The French families would close their shutters as soon as the lights went on in their homes (Møhl, 1997). The lit home at night was a private space, separate from the public sphere. Sending light into the streets was considered 'showy', and risked accusations of inviting attention. By contrast, the Indian residents would install their first domestic lights in the porches at the front of their houses, rather than inside (Kumar, 2015a). The light was a sign of openness, hospitality, and putting it in front of the house, in effect offering illumination to the public sphere, was understood as an act of generosity. So the home at night can be both an inward- and an outward-reaching space, but the outward-reaching aspects have multiplied as homes have become more connected. This interconnection can be understood negatively – as part of the pressure to answer emails and maintain social connections throughout the day and night – though it can also help to connect people who are trapped in the home with the wider world. However, even where the home is interconnected, there is typically still control over who physically enters the domestic space, as public ties are severed at the edge of the property.

This severance of ties fosters intimacy with those who are allowed inside. Ekrich (2006, p. 161) explains that 'night was a fertile time for romantic liaisons of all sorts, notwithstanding rigorous restrictions against sexual activity,' in early modern times. The hours of darkness provided a cover for sexual liaisons that would have been forbidden during the day. The celebrated balcony scene in Shakespeare's *Romeo and Juliet* reflects this: it takes place at night so the lovers can avoid the surveillance of their respective families. Juliet's exclamation 'What man art thou that, thus bescreen'd in night / So stumblest on my counsel?' reveals that merely approaching someone at night, particularly in the context of a potentially romantic relationship, was to engage in an act of intimacy. The wonderfully evocative seventeenth-century word 'bundling' describes the custom of allowing courting couples to spend the night together in the home of the woman's parents (Ekrich, 2006). While sexual congress was forbidden, and indeed apparently rare, during bundling (Fischer-Yinon, 2002), the practice allowed couples to develop the intimacy that comes with spending the night together. In other words, it was the simple act of spending time in close proximity to each other in the home at night, rather than sex, which cultivated intimacy. Therefore, we may say that the domestic night has long been central to the production of intimate relationships.

The binding that intimacy creates is in part social – a symptom of the rhythmic night which intensifies the social relations that we have with those who are still around us at times when our connections to the rest of society are reduced. However, we are also connected to others through our experience of darkness, which has become a topic of social science research at a time when it is facing an existential threat (Edensor, 2017). As noted in Chapter 3, interest in spaces of darkness in the public sphere has increased since light pollution has altered our relationship with the sky. However, darkness has an affective set of relations, too. Drawing on phenomenological theory, it seems that the experience of darkness in the home has a particular set of characteristics that not only increase intimacy but also play a crucial role in the night's production of the domestic space.

Merleau-Ponty's discussion of darkness is brief, but his observations laid the foundations for later phenomenological work. He argues that 'night is not an object before me; it enwraps me and infiltrates through all my sense, stifling my recollections and almost destroying my personal identity' (Merleau-Ponty, 1962, p. 283). In other words, the night penetrates into our sense of self and reinforces the porosity of the body. Merleau-Ponty suggests that this happens because darkness removes the safe perceptual field of daytime: that is, it is the experience of darkness which 'enwraps and infiltrates', rather than night itself. Minkowski was one of Merleau-Ponty's students and added significantly to his work by focusing specifically on darkness. He was especially interested in the relationship between spatial and temporal disorientation: 'pathological disorientation in time is accompanied by a disorientation in space, as if the two disorientations were only expressions of the same disorder' (Minkowski, 1970, p. 13). Minkowski saw disorientation increasing in the modern era, and he used his research into pathological experiences to try to recapture 'lived' time and space more broadly. For him, this timespace is not dominated by technology or capitalism, but by the experiences of the open and corporeal subject; and the limitations of conceptions of bounded subjectivity are revealed during periods of darkness.

Minkowski, then, views darkness as more than the mere absence of light. Rather, he feels it is defined by the shrinking of the bounded, perceiving space of the body. In darkness, our perceptual field shrinks and the boundaries of the 'detached' body implode to such an extent that they may no longer exist. This allows darkness to 'engulf' or 'penetrate' the self. He argues that darkness

> does not spread out before me but touches me directly, envelops me, embraces me, even penetrates me, completely, passes through me, so that one could almost say that while the ego is permeable by darkness it is not permeable by light. The ego does not affirm itself in relation to darkness but becomes confused by it.
>
> (Minkowski, 1970, p. 429)

For phenomenologists, the self is made by recognizing 'others': by understanding that 'I' am different to 'you'. In darkness, however, this is impossible, and the affirmation of the ego cannot happen in the same way. Rather, our sense of self is revealed to be permeable – not detached from the world around us, but subject to a wider range of inputs and flows than we typically believe. This experience of darkness reveals that our bodies were never detached or bounded in the first place – as I have argued in this book. Rather, our subjectivity was always embodied, and that embodiment was always porous. In other words, in light, we are able to recognize ourselves as selves by seeing ourselves as separate from other objects. By contrast, in darkness, we are unable to do this as we cannot affirm the differences between ourselves and the objects around us.

This openness to others means that, in darkness, relations will be inherently more intimate. Even in the contemporary home, moving between varying levels of artificial light and shadow will produce relationships which are more intimate than those in the day. As light is the mechanism through which some other appears to me, in darkness I am rendered open to that other. The perceptual field created by light is thus reconceived as a protective field – a boundary that separates me from other objects which are 'held' at a distance. In night, however, these objects are no longer held at a distance; instead, they are hidden and unknown. Cataldi (1993, p. 48) describes this as 'spatiality without things' – a time when the difference between 'things' is removed. Our sense of space is clear in the protecting 'bubble' of light, whereas in the dark we have a more fluid spatiality, one in which the internal and the external more readily intermix. This renders the self more open to other objects, so it may be more readily affected. This facilitates intimacy but also vulnerability. In other words, openness may have either positive or negative outcomes (or a combination of the two). To be intimate is to be vulnerable, and the openness of darkness is one of the key mechanisms through which the home at night can be both a space for facilitating personal ties and a space of danger and fear.

Home: at the edge of urban and beyond

Despite the increased connectivity of home at night, this timespace remains one in which intimacy is created and our subjectivity is vulnerable. This opens up three ways in which the home at night remains distinct and separate from the urban day. I will argue that these fundamental aspects of the domestic night run counter to the defining characteristics of urban life, which suggests that the 'planetary urbanism' of global society does not constantly penetrate into the home at night.

First, I want to highlight the danger, fear and isolation that may be felt in the home at night due to a sense of vulnerability. Sociologists speak of 'social isolation' (Dean, 1961), particularly in the context of health-related research. This emerges from a 'lack of social interaction or lack of social integration' (Stets, 1991, p. 670), and it is often diagnosed as a long-term problem. However, many people experience what amounts to recurring temporal social isolation at night. This topic has not been studied extensively, but there are snippets in the literature which hint at the underlying nocturnality of loneliness. For example, Australian researchers found that 'some older people indicated that they experience loneliness most keenly at night even if the daytime is full and satisfying' (Stanley *et al.*, 2010, p. 411), while loneliness in relation to grief is reportedly worse at night (Bennet and Victor, 2012, p. 42). Moustakas (1961) writes powerfully about the intensity of loneliness in hospitals at night, when there are no visitors and the activities of the day typically cease. In more extreme social settings, the presence of night-time curfews in conflict zones can trap people in their homes at night (Malmström, 2014, p. 28). While the connected home mitigates against this, as I argued in Chapter 3, even those forms of connectivity are reduced to the minimum infrastructural provision. So forms of temporal isolation and loneliness persist, which can make the home at night a timespace of danger and fear.

Fear in the home at night extends to domestic violence, too. Studies have noted that much of this, both physical and mental, takes place at night, and it has been suggested that abusers use the night to facilitate their abuse. In a study of the timing and location of rapes in Christchurch, New Zealand, for example, Pawson and Banks (1993, p. 58) found that around 75 per cent of attacks took place between 21.00 and 06.00, and around 60 per cent occurred in the home; in other words, 'most of the recorded attacks occurred at night, during or immediately after hours of social contact, or for those alone, when community surveillance is minimal'. More recently, research among domestic violence victims in Scotland has revealed that the abuse is perpetuated through the creation of 'chronic fear which builds up over the long term and leads to significant trauma and negative effects on health and wellbeing' (Pain and Scottish Women's Aid, 2012, p. 6). This long-term fear has some of its foundations in the night. For example, several of the participants in the Scottish study highlight the devastating impact of nightmares, such as: 'the flashbacks you get as well ... that's what I couldnae cope with and [I was] terrified at night' (p. 26). It is when darkness falls that the mental effects of this form of abuse are most keenly felt. The removal of safety and security from the home at night comprises the removal of a timespace in which people have the opportunity to renew and replenish their sense of self and security. Night also binds people closer to those with whom they live: for example, Lowe et al. (2007) point out that night often forces victims to sleep alongside their abusers. They found that women in abusive relationships adopted a variety of strategies to cope with this: for instance, they might avoid sleep altogether while the partner is at home; or they might remain awake until the partner falls asleep. Such routines create a 'connection between sleep deprivation and the establishment of a regime of power and control by one person over another – the hallmark of domestic violence' (Lowe et al., 2007, p. 558). Here, the mere proximity of two bodies in a darkened space allows the perpetrator to exercise power over the victim. Night seems to be a time when domestic violence is more prevalent, and in which its effects are more strongly felt.

Second, I wish to emphasize the relative quiet of most homes at night. Notwithstanding the narratives related by many – including myself – the expansion and domination of capital, control and the day are not yet complete. In the UK alone, four million homes are not connected to the gas network, and thousands of others are not on the national electricity grid or served by broadband networks. Meanwhile, in informal settlements across the globe, access to various grid facilities is partial, precarious or non-existent. Many rural homes are similarly unconnected. In other words, especially at night, these homes operate without the connections that define planetary urbanization. This has several impacts for the domestic night. Lighting and heating regain their materiality, difficulty and danger as they are associated with coal, wood, tallow and other highly flammable materials. Social connections at night are less common, too. The aforementioned studies by Kumar (2015b) and Møhl (1997) reveal the ways in which the quiet of night impacts unevenly. For instance, Kumar found that Indian women remain in the darkened and disconnected parts of the home, while the men spend time in the lit areas. Both researchers feel that there are aspects of the domestic night which remain defiantly uncolonized, even after the introduction of some infrastructure, with homes still largely disconnected from the networks that define the day.

Third, it is worth noting the various forms of care, exchange and intimacy that still fall outside dominant capitalist relations. The home is 'usually understood to provide a refuge from global capital and the implications of globalisation' (Valentine, 2008, p. 2103). Even if we accept that homes and associated social reproduction feed into global capital, they contain many relations and practices that occur outside of it; in other words, the home and the ties of intimacy, friendship and family that are associated with it remain, as they have always been, on the borderlands of capital. Moreover, as the combination of social and biogeoastronomical night makes daytime expansion difficult, the home at night has the potential to move even further from the capitalism of the urban day and become a timespace for fostering resistance or practising alternative ways of living (Jarvis, 2011).

In this section I have articulated three ways in which the domestic night disturbs narratives of planetary urbanism. It does so in quite ambivalent ways: from fostering intimacy that has the potential to challenge power to facilitating social isolation and abusive relationships. As an area of academic research, the home at night remains rather underexplored, but here it pushes us beyond the normal tropes of night-time research, which has typically explored city streets, infrastructure and businesses. It also draws us into timespaces which are located within the built environment of the city, yet remain peripheral to or fully outside the relations of capital and mobility that seem to define urban life. The domestic night is about isolation, lack of services and darkness. By contrast, in theories of urban life, the city is defined by a number of characteristics – density and diversity of population, affects of pressure, capitalist exchange, increased surveillance, proximity to the state – that are either absent in the night-time home or only just beginning to spread into parts of it. In other words, the domestic night appears to reside on the very edge of – or even outside – the conditions of living in the city, and as such remains outside the process of planetary urbanization.

References

Bennet, K.M. and Victor, C. (2012) '"He wasn't in that chair": What loneliness means to widowed older people', *International Journal of Ageing and Later Life*, 7(1), pp. 33–52.
Berardi, F. (2009) *The Soul at Work*. London: Semiotext(e).
Berlant, L. (2011) *Cruel Optimism*. Durham, NC: Duke University Press.
Bille, M. (2015) 'Lighting up cosy atmospheres in Denmark', *Emotion, Space and Society*, 15(1), pp. 56–63.
Bille, M. (2017) 'Ecstatic things: The power of light in shaping Bedouin homes', *Home Cultures*, 14(1), pp. 25–49.
Bille, M. and Sørensen, T.F. (2007) 'An anthropology of luminosity: The agency of light', *Journal of Material Culture*, 12(3), pp. 263–284.
Blunt, A. (2005) 'Cultural geography: Cultural geographies of home', *Progress in Human Geography*, 29(4), pp. 505–515.

Buse, C. and Twigg, J. (2014) 'Women with dementia and their handbags: Negotiating identity, privacy and "home" through material culture', *Journal of Aging Studies*, 30, pp. 14–22.
Butler, J. (2015) *Notes toward a Performative Theory of Assembly*. Cambridge, MA: Harvard University Press.
Cataldi, S.L. (1993) *Emotion, Depth, and Flesh: A Study of Sensitive Spaceonty's Philosophy of Embodiment*. Albany: State University of New York Press.
Dean, D.G. (1961) 'Alienation: Its meaning and measurement', *American Sociological Review*, 26(5), pp. 753–758.
Domosh, M. (1998) 'Geography and gender: Home, again?', *Progress in Human Geography*, 22(2), pp. 276–282.
Edensor, T. (2017) *From Light to Dark*. Minneappolis: University of Minnesota Press.
Ekrich, A.R. (2006) *At Day's Close*. London: Phoenix.
Fischer-Yinon, Y. (2002) 'The original bundlers: Boaz and Ruth, and seventeenth-century english courtship practices', *Journal of Social History*, 35(3), pp. 683–705.
Hardt, M. and Negri, A. (2006) *Multitude: War and Democracy in the Age of Empire*. London: Penguin.
Harrison, P. (2009) 'In the absence of practice', *Environment and Planning D: Society and Space*, 27(6), pp. 987–1009.
Head, E. (2005) 'The captive mother? The place of home in the lives of lone mothers', *Sociological Research Online*, 10(3): www.socresonline.org.uk/10/3/head.html (accessed 21 October 2017).
Healey, N. (2016) 'How effective is your sleep tracker?', CNET, 16 February: www.cnet.com/news/how-effective-is-your-sleep-tracker/ (accessed 21 October 2017).
Jarvis, H. (2011) 'Saving space, sharing time: Integrated infrastructures of daily life in cohousing', *Environment and Planning A*, 43(3), pp. 560–577.
Kumar, A. (2015a) 'Cultures of lights', *Geoforum*, 65, pp. 59–68.
Kumar, A. (2015b) 'Energy access in an era of low carbon transitions: Politicising energy for development projects in India'. Unpublished Ph.D. thesis, Durham University.
Lader, D., Short, S. and Gershuny, J. (2006) *The Time Use Survey, 2005*. London: Office for National Statistics.
Lowe, P., Humphreys, C. and Williams, S.J. (2007) 'Night terrors: Women's experiences of (not) sleeping where there is domestic violence', *Violence against Women*, 13(6), pp. 549–561.
Maille-Petty, M. (2010) 'Edge of danger: Electric light and the negotiation of public and private domestic space in Philip Johnson's glass and guest houses', *Interiors*, 1(3), pp. 197–218.
Malmström, M.F. (2014) 'The sound of silence in Cairo', *Anthropology Now*, 6(2), pp. 23–34.
Merleau-Ponty, M. (1962) *The Phenomenology of Perception*. London: Routledge.
Minkowski, E. (1970) *Lived Time*. Translated by Metzel, N. Evanston: Northwestern University Press.
Møhl, P. (1997) *Village Voices: Coexistence and Communication in a Rural Community in Central France*. Copenhagen: Museum Tusculanum Press.
Moustakas, C.E. (1961) *Loneliness*. Englewood Cliffs: Prentice-Hall.
Pain, R. and Scottish Women's Aid (2012) *Everyday Terrorism: How Fear Works in Domestic Abuse*: www.scottishwomensaid.org.uk/files/EverydayTerrorismReport.pdf (accessed 8 August 2017).

Pawson, E. and Banks, G. (1993) 'Rape and fear in a New Zealand city', *Area*, 25(1), pp. 55–63.
Pink, S., Leder Mackley, K. and Moroşanu, R. (2015) 'Researching in atmospheres: Video and the "feel" of the mundane', *Visual Communication*, 14(3), pp. 351–369.
Powells, G., Bulkeley, H., Bell, S. and Judson, E. (2014) 'Peak electricity demand and the flexibility of everyday life', *Geoforum*, 55, pp. 43–52.
Sagan, C. (1994) *Pale Blue Dot: A Vision of the Human Future in Space*. New York: Random House.
Stanley, M., Moyle, W., Ballantyne, A., Jaworski, K., Corlis, M., Oxlade, D., Stoll, A. and Young, B. (2010) '"Nowadays you don't even see your neighbours": Loneliness in the everyday lives of older Australians', *Health and Social Care in the Community*, 18(4), pp. 407–414.
Stets, J.E. (1991) 'Cohabiting and marital aggression: The role of social isolation', *Journal of Marriage and Family*, 53(3), pp. 669–680.
Swyngedouw, E. (1997) 'Power, nature, and the city: The conquest of water and the political ecology of urbanization in Guayaquil, Ecuador: 1880–1990', *Environment and Planning A*, 29(2), pp. 311–332.
Tolia-Kelly, D. (2004) 'Locating processes of identification: Studying the precipitates of re-memory through artefacts in the British Asian home', *Transactions of the Institute of British Geographers*, 29(3), pp. 314–329.
Valentine, G. (2008) 'The ties that bind: Towards geographies of intimacy', *Geography Compass*, 2(6), pp. 2097–2110.
Virno, P. (2004) *A Grammar of the Multitude*. Los Angeles: Semiotext(e).

7 Towards nightology and the temporal limits to urbanism

When I started writing the proposal for this book and giving presentations of ideas to refine my argument, before then writing the manuscript, one of my major aims was to explore questions about the night that intrigued me. In particular, two problems occurred to me. The first related to assertions that night is a time outside of 'everyday' social conventions and restrictions: is the night really the sort of emancipatory timespace that several writers (Self, 2014; Edensor, 2015; Dunn, 2016; Stone, 2017) have described? If it is, how do we consider contrary evidence that the night is also a timespace into which people are forced to enter in order to seek work or escape, and that, as a timespace, it is one in which oppressed groups are more vulnerable than those who are in positions of power (Melbin, 1987; Valentine, 1989; Patel, 2006; Crary, 2013)? Second, what is the urban night's relation to the (planetary) urbanism of day? Can it be understood as part of the same system even though empirical research suggests that it is often radically different?

I hoped that writing this book would help me answer these questions; it seemed as good a reason as any to write a book. Of course, it was unreasonable to expect a simple yes/no answer. The more we explore them, the more the concepts associated with night are revealed to be multiple and contradictory. In the Introduction I suggested that night might be a 'haecceity' (Deleuze and Guattari, 1987, p. 287); as such, it cannot have a single narrative unifying all of its elements. Moreover, such a narrative would not be particularly desirable anyway. Instead of providing an answer to these questions, then, this book has attempted to illustrate the multiple ways in which some of the diverse forms of living at night push at the common understandings of urban life.

Before summarizing the arguments, I want to reiterate the case for why this matters. Theorists of planetary urbanism have had a major influence in urban geography in recent years. One of the core contestations is the positing of urban theory 'without an outside' (Brenner, 2014), a claim which has been criticized from a variety of perspectives. Scholars have argued that it falls foul of a (masculinist) tendency towards universalization (Oswin, 2016); that it overlooks the realities of urban agglomeration (Storper and Scott, 2016); and that it is Western-centric (Derickson, 2015). However, many others have found the claims of planetary urbanization useful, and 'softer' variants of it abound in social science, literature and wider public discourse. For their part, the authors of such claims have made a range of interesting defences of their assertion of an urbanized world. In particular, they argue that their analysis captures the spread and depth of urban capitalism, and that claims to the contrary advocate empirical particularism over theoretical complexity (Smith, 2013). Here, the academic importance of this debate relates to our conceptualization of cities but also to wider considerations of epistemology and ontology. In other words, the differing ways in which cities are described and understood are important because they turn on our relationship to evidence, theory, analysis and empirics. What do we consider good evidence, good analysis and good theory? These issues are all up for grabs in this debate.

Beyond this, however, the discussion of the limits to the city can also influence how urban problems are managed. If we take the city as equivalent to the globe, then environmental, infrastructural and population problems must be imagined at the urban scale. At the same time, issues of subjectivity, socio-economic inequality and oppression need to be explored on a planetary basis. By contrast, if we continue with what we might call an 'agglomeration' model, which sees cities as distinct geographical nodes within systems, then that has its own series of implications, reinscribing a focus on the particular problems of individual cities as centres of concentrated human life and political and/or economic power.

Planetary urbanization: many cities, many planets

In Chapter 1, I posited a definition of the urban as processual, drawing on Brenner and Schmid's (2015, p. 172) arguments that cities are a *materialization* of industrial capitalism into particular places and its *transformation* into 'concrete, temporarily stabilized configurations of socioeconomic life, socio-environmental organization and regulatory management'. Developing this, I argued that we needed to expand Brenner and Schmid's definition in two ways. It requires, first, a deeper role for the agency of the ecological and, second, a larger role for the subjectivities which produce and are produced by urban life. These both respond to the two elements of Guattari's *three ecologies* that planetary urbanization has covered less strongly. They also point us towards small-scale interactions, or perhaps the ways in which a focus on the planetary demands that we consider how small- and large-scale interactions intersect. The intervening chapters have explored the ways in which the city at night has grown and become an essence of 'urbanism', but they have also identified a series of moments that do not fit easily with this view of global urban life.

One of this book's key arguments is that there is more than one way in which urbanization can be planetary. In particular, *planetary* urbanization needs to be understood as distinct from *global* urbanization. Of course, there are many ways in which the urban subjectivity machine is both planetary *and* global. We can see in the spread of artificial lighting ways in which *planetary* being has become subject to the logics of global urbanism. Power lines and street lights follow transport networks, producing lines of lighting across the globe that have ecological impacts in a multitude of ways. City landscapes become 'spectacular', lit with a fascinating array of lighting technologies. Through this infrastructure, the planet and its ecologies become somewhat urbanized as light pollution transforms animal behaviour and changes prey–predator relations. However, we also find some aspects of planetary being which seem to reside on the edge of – or even outside – the urbanized ways of living that we have identified. Experiencing darkness alongside rest and sleep at night is one of the ways in which humans have a planetary being. I have termed this the 'biogeoastronomical' night, as it emerges out of biological adaptations to geographical and astronomical features. The experiences of darkness in the home that occur in the biogeoastronomical night, as I argued in Chapter 6, are at the core of the production of subjectivities. Such experiences can also take place in the uncanny experience of wandering alone through streets at night, and here they can be moments of rupture of non-urban planetary being. In other words, while planetary, they do not seem to be moments of *urban* life.

With regards to *global* urbanization, we can see multiple other limitations to the earth's urbanization. For instance, when looking at the globe at night, we see many gaps in the spread of artificial lighting (Pritchard, 2017). Urban society may be spreading, and its influence spreading even further, but at night there are still plenty of dark spaces. Even though the protection provided by Dark Skies Parks ties these reserves into global systems of capital, they seem to preclude the introduction of an 'urban way of life' within them. Furthermore, as our discussion of the night-time economy explored, cities at night have not undergone anything like the homogenization of urban forms that has occurred in daytime cities, despite the pressures of place marketing (see Chapter 5). Thus, nocturnal urban life forms much less of a coherent global system than daytime urban life.

So, night complicates the picture of an increasingly urban world by revealing the multiple ways in which cities can be planetary, and the multiple ways in which the planetary might not be urban. Whether they have centred on 'planetary urbanization' or more hesitant descriptions of 'urban worlds', the large-scale debates that have permeated urban studies over the last decade have largely overlooked questions of the earth and temporality in favour of questions of the globe and space. By focusing on the night, I hope this book has revealed some of the ways in which the boundaries to urbanism as a way of life persist.

Three narratives of the urban–world–night relationship

In part, the preceding chapters have been concerned with narratives of the night-time city. Crucially, while identifying the night-time city as multiple and eluding a single narrative, there are multiple interlocking, convincing narratives that capture many of the ways in which it coheres into a single unit. To delve a little further into the night itself, and inspired by Gwiazdzinski's (2005) attempt to outline night-time planetary futures, below I present three narratives of the urban–world–night relationship. These are not definitive, and they are generalizations. However, I trust they will clarify some of the key trends in the urban night that this book has explored.

Narrative 1: A time-globalized world: the city and world as one

I have written of light pollution which threatens habitats, cities that have vibrant night markets, and homes that are connected to others across the globe without any requirement to switch on or off. This continued 'colonization of the night' (Melbin, 1987) could be understood in terms of the frameworks offered by writers on the night such as Gwiazdzinski (2005) and Crary (2013), both of whom have emphasized the continued expansion of day into night. Just as Melbin identifies a similarity between colonial expansion over space and social expansion over time, we can start to describe a temporal version of globalization in which the expansionary features of capital that have reworked space and place are inscribed onto time. Such tendencies are embedded not just in capital but also in socio-political expansion and the rise of mobility, information and communications technology. In a narrative that stretches across multiple empirical phenomena and multiple domains of human activity, night has been lost or is in the process of being lost. In this formulation, the 'urban age', as it has been labelled (Brenner and Schmid, 2014), is one in which activity is continuous, in which the city and urban life more broadly are already – or are becoming – incessant.

Incessancy should be understood as the continuous operation of life. It is not without rhythm, or without quieter moments, but rather a situation in which:

1 Spaces throughout the city and its rural hinterland are active, in some ways, across the twenty-four hours of each day.
2 Multiple areas of public and domestic life continue operating across twenty-four hours.

The existence of incessancy is one of the prerequisites for planetary urbanism. As I have argued, the presence of activity through the night is one of the core differences between urban and rural living, and the increased activity of infrastructure around the globe at night, along with the circulation of an aesthetics of night, provides further evidence of the ways in which urban life has globalized.

We see this form of the city in the public spaces of many global cities, and in the always-on spaces of global mobilities (airports, control centres and so forth). It fits with much of the empirical evidence that I have presented throughout this book. Researchers have repeatedly discovered ways in which the night has been 'lost' – a term that appears in several texts (Ekrich, 2006;

Pottharst and Könecke, 2013). Pottharst and Könecke, in particular, discuss several losses: they assert that darkness, health, sleep and heritage have all disappeared in the face of a rampant expansionary day in the form of the emergent – or indeed emerged – 24/7 city. This loss of temporal diversity mirrors the loss of diversity in cultures, languages and ways of living that have been witnessed on the spatial scale. In other words, we might posit a 'diurnalization' of time alongside a 'Westernization' or 'McDonaldization' of the physical world. In city centres, we can now identify certain places and moments which seem to be as busy at night as they are in the day, if not busier. Underlying these and the city more broadly are networks and infrastructures which continue to operate at night, and again I turn to the use of night-time lighting to illustrate the broader concept of planetary urbanization and as evidence of the city that continues to operate throughout the night (Brenner, 2013, p. 81). Furthermore, we can draw parallels between the anti-globalization movement and attempts to counter the spread of lighting pollution. Groups such as the International Dark Sky Association assume some of the same roles as, for example, the 'slow food' movement or Greenpeace: they seek to protect natural and traditional ways of living against the forces of global expansion. Equating temporal expansion with spatial expansion is useful as it highlights some of the ongoing changes in self–city–world relationships. Moreover, from an ecological perspective, this notion emphasizes the connections between the damaging effects of globalization and the damaging effects of the infrastructures that are required to keep cities operating throughout the night.

So, if we view the city as a subjectivity machine – in other words, if we 'treat the mass/aggregate of urban and architectural machinery as machinic components, all the way down to their smallest subgroupings' (Guattari, 1993, p. 145) – then in the expanded/exploded city we can see a conceptualization of the urban which responds to the increased diversity of actors who produce subjectivity. Complex infrastructural systems, such as the electricity network, must be understood in terms of the maintenance workers who keep them running, the computer programs that timetable lighting sequences, the electrons which power them and the night sky whose natural illumination they obscure. Building on the Marxist insights of planetary urbanism, specifically the tendency towards expansion, equalization and differentiation (Smith, 1984), we see a temporal process which mimics the spatial expansion of capital. This compelling tale of the 'diurnalized city' relates to both a city that is globalized and a form of urban living that is inherently planetary. It describes dominant tendencies in urban life and fits with many of the processes described in this book. Ever more of the world at night is becoming more like the day.

However, lines of flight remain; deterritorializations will be inevitable. In this chapter, I have already argued that the night seems to fall outside the planetary at key moments. As Oswin (2016, p. 5) argues in relation to claims of a universal planetary urbanism, 'Yes, capitalism is everywhere. But so is everything else. And there are outsides – constitutive ones – all over the place.' I support this assertion; and, as I go on to explore in the next two narratives of the urban night, we can easily identify nocturnal people, places and experiences which seem to fall outside the urban world, or at least form its borders.

Narrative 2: The fragmenting frontier and the night

As numerous writers have argued, night has always been more complex than simple narratives allow. Anthropologists working with traditional and nomadic communities have revealed traditions and customs that extend well into the night, including such activities as hunting and fishing, rituals and performances. Historians have documented a wide range of people working in, walking through and inhabiting public spaces at night. The early era of artificial illumination witnessed a variety of nocturnal responses, including but not limited to: 'Leveller'-style groups contesting the spread of light into the dark; the emergence of alternative popular cultures in cities for people who craved lifestyles outside the surveillance of the day; and a series of jobs shifting into the night, some out of necessity and others to exploit the extended operating hours of machinery. Night has never been the empty time that narratives of expansion, frontiers and colonization have presented.

Yet, if night was never an empty time, it is not now becoming a time of ever-present, incessant activity. Rather, the frontier night is undergoing transformations in relation to the expansionary forces of the day and losing some of its cohesion as a result, but it is not disappearing. Perhaps the night was a coherent frontier at one time, a timespace where the bold and the brave joined those who had been evicted from the day to form a society that was somewhat independent and separate from the norm. We could even identify a specific period of time – perhaps from the first artificial illumination of the late eighteenth century until the emergence of global night-time connections between cities with enhanced telecommunications in the mid-twentieth century – when the night in many cities could be understood as an expansionary space at the edge of – but not a full part of – a coherent but limited urban system. It is this period that Melbin (1987) describes in *Night as Frontier*. Moving beyond his frontier, though, we can understand this night as consisting of contradictions, encounters and the meeting of multiple cultures. As these 'contact zones' evolve, the clashes can fragment, spreading from the frontiers and into the wider 'imperial' society (Morrissey, 2005). We can see some of this process in the contemporary fragmentation of the night. Spaces within the night itself have become more 'daylike': night-time markets, spaces of infrastructural maintenance, the night sky and the bars and restaurants of the night-time economy are all brighter, more active and more open to the activity of capital than ever before. However, other spaces have become more disconnected, more distant, as darkness has returned to cities due to the demands of sustainability, for example. We can also see the ways of being in the night stretching into the day, for instance in the night-time call centres that handle daytime calls from other time zones, or the pubs and bars that open for breakfast. This is captured by the concept of the fragmenting frontier – a borderland which is not as coherent as it once was.

Concepts of fragmentation and heterogeneity lend themselves to notions of assemblage and machinic subjectivity, as previously established. The vision of boundary and frontier presented in postcolonial theories – as 'a fringe, as a vague intermediate state or landscape or ... region' (Naum, 2010, p. 101) – speaks to the broader understandings of the city that have been developed in urban studies over recent years. If the city is a process, or a series of ways of living, or a subjectivity machine, then it must have areas which are vaguer or more liminal. I have previously advocated a position along these lines (Shaw, 2015), and the concept of the fragmenting frontier remains, for me, an apt framework for discussing many urban spaces at night, whereby the boundary zone between urban and non-urban is becoming increasingly complex yet not disappearing. There are, however, limitations to this idea. Most obviously, if frontiers are always to be understood as partial, heterogeneous contact zones, then what does it mean to say that something which is partial and limited in the first place is now fragmenting? At what point does heterogeneity become fragmentation? It is difficult to establish the precise moment when this happens. Nonetheless, given that night is a haecceity, a fluid and intangible assemblage, we should not take an inability to determine its boundaries as evidence for it not existing. The advantage of the idea of fragmentation is that it offers a means of keeping the sense of a frontier in place even when some elements of the night-time city are becoming more like day. Nevertheless, we might go further ...

Narrative 3: Outside city walls: night and nature beyond urban life

For some of the advocates of planetary urbanism, little – perhaps nothing – falls outside of the urban (Merrifield, 2012). Merrifield's version of planetary urbanism evokes the idea of the urban as immanent substance, always with us, gaining expression in particular forms or 'attributes', such as the built environment, an encounter with the 'other' or forms of urban governance. He feels it is not so much the world that has been urbanized, but life and the possibility of living. Thinking through the urban and the world together, and through this vision of an urban lifestyle, we see the potential for limitless urbanism. But what if we understand the night as something *outside* the urban? Returning to Merrifield, many of his urban attributes are reduced or even absent at night: forms of governance cease; encounters become less common and people may actively avoid public spaces due to safety concerns; even the built environment changes, in our perception but also physically, due to curfews or locked gates, shadows and artificial lighting, lower temperatures and other sensed spatial encounters.

As well as the earlier discussion about the multiple ways of being planetary, a few moments might just fall outside of urban life. The first of two I shall consider here entails looking up at the night sky and seeing the Milky Way. This experience challenges narratives of planetary urbanism in two ways. First, it is distinctly non-urban as it relies on none of the technologies of city life; indeed, even the most basic of urban forms may preclude it. It has been accessible to humans since well before cities were first founded. It is not predicated on encounters with other people; certainly, it can be shared, but it can also be experienced alone. It has no relation to forms of urban governance, and it does not depend on any relationship with capital (although, as Chapter 3 discusses, the development of Dark Sky Reserves and the associated tourism are starting to subject this view of the night sky to both regulation and the rules of capital). Going further, these elements are not just non-urban but actively *anti*-urban: the view of the night sky is better and more accessible outside cities. Second, seeing the Milky Way is also not entirely planetary: rather, it is *inter*-planetary, or even interstellar. This may be a slightly pedantic point, but it seems a suitable response to planetary urbanism's totalizing claims. If we can – in a distant, removed, but still very real way – perceive the light of distant stars, this reminds us that the earth is not a complete whole, not a system that is independent of the wider universe without any inputs or outputs. Rather, planetary being is simultaneously interplanetary being (Kearnes and van Dooren, 2017). It does not stop at the (diffuse) borders of the earth's atmosphere.

If the view of the Milky Way illustrates a nocturnal experience that falls outside of planetary urbanism on a large scale, then the domestic night illustrates the night outside of planetary urbanism on a small scale. Alone in a darkened room at night, it is questionable to say whether 'the urban' has much meaning: the occupant of the room cannot access services which are shut down, social networks which no longer operate, or the 'buzz' of urban life. They will slide between states of dreaming and wakefulness as people always have, entering moments when they are almost, if not entirely, outside the social (Harrison, 2009). This suggestion is perhaps more tentative than that of viewing the Milky Way; after all, in many parts of the world, the nocturnal home is heated by urban infrastructure and furnished with items purchased in the city. Nevertheless, this does not alter the fundamental interplay of light and dark that is experienced at night when drifting between wakefulness and sleep. This may be planetary, biogeoastronomical being, but we may question the extent to which it is truly urban.

Towards nightology

In these three narratives of the urban night, we have seen a variety of ways in which night can inform us about urban life. Taking an 'ecosophical' approach to the city, focusing on the production of subjectivity, environment and society together, can help us to grasp the multiple dimensions of urban life that become increasingly interconnected as urban systems expand and diversify. Night, in particular, encourages us to consider the fundamental conditions of the planet and society that shape urban life. With this in mind, I wish to end with four further directions for 'nightology', and indeed hope to make the case for the night as a topic of study in and of itself.

Nights beyond urban public space

Chapter 6 bemoaned the paucity of night-time research outside of urban public spaces. While the themes of the night-time economy and public lighting have become common areas of research, spaces such as the home, offices and workplaces, community centres and other civic spaces are much less understood. Moving further, there has been minimal research into rural spaces at night, including villages, agricultural land and so forth. Organizations and other institutions could also be the focus of more nocturnal research, with interesting questions to be asked about the continuity between days and nights. As I have argued in this book, the production of urban life does not occur exclusively in the public spaces of the city, so night-time research has a long way to go to match the diversity of spaces that have been studied during the day.

Diverse locations and experiences

Relatedly, most of the existing research is Eurocentric. In this book I have attempted to insert some examples from outside the prevailing Euro-American focus of academic geography, but both my own experience and the published literature inevitably lead towards a small number of cases. Nocturnal research has been heavily biased towards Western Europe, the USA and Australia. The merits of engaging with a wider range of night-time cities and expanding the palette of nocturnal research should be obvious. Exploring a wide diversity of night-time experiences would improve our understanding of what the night is ... and what it could be.

Interdisciplinary

Though my background is in geography, I have tried to incorporate research from cognate disciplines. Nocturnal research already cuts across traditional disciplinary divisions, but we could do more to foster this. My focus has largely been on social science, but wider areas of knowledge could be explored, too. Psychological and biological research has investigated the impact of night-shift working on individuals and their relationships, but few efforts have been made to integrate this into urban studies or other social sciences. However, the interdisciplinary 'Dark Side of Light' research project has been exemplary in connecting to ecological research on light pollution (Hölker *et al.*, 2009), and similar collaborations would certainly enhance the field. The literature on lighting – a key area of research – seems divided between the studies of engineers and those of lighting designers, with the latter oriented towards art and aesthetic practice, and the former towards scientific and technological precision. Clearly, more dialogue between the two groups would be beneficial.

Indeed, as it is a field that combines the natural and the social, there is no reason for nocturnal research to be anything other than inter- and cross-disciplinary.

Conceptualizing night

In this book, I have attempted to conceptualize night in a variety of ways. In particular, I have focused on two interconnecting forms of night: the biogeoastronomical and the socio-cultural. I adopted this approach in the hope of unpacking the different ways in which night seems to operate. In particular, I wanted to distinguish between the effects created by, on the one hand, the naturally occurring rhythms of light and dark, plus the biologically evolved responses to these rhythms, and, on the other, the socially constructed and contingent practices and meanings that have emerged around the biogeoastronomical night. In practice, these two elements intersect with each other, but I feel that the conceptualization helps to establish exactly what night is. It would be beneficial, though, to combine it with more diverse empirical research in order to extend the range of understandings of night.

In a poem titled 'Do not go gentle into that good night', Dylan Thomas (1966) equates night with death – as both should be resisted strongly – and light with living. The night has long been perceived as the 'other', outside of the norm, outside of what counts as normal society: 'night is physically the same as any other time, except that it is dark' (Melbin, 1987, p. 1). In this analysis, I have offered a vision of how our picture of both city and world changes when we put the night at the centre of our analysis. There are, as I have just suggested, many more routes that such an approach could take. I end this book with Thomas's poem, though, in the hope of encouraging others to follow these routes *gently*. Research into the nocturnal city could help us to negotiate some of the complex debates surrounding claims of planetary urbanization, but to do so we should be tentative in our moves to theorization, and should remain grounded in both literature and empirics. The night-time city is at once an intensified urban form of living and a timespace in which the city loses many of its inherent characteristics. This paradox ought to provide fruitful areas of research for anyone looking to understand urban life.

References

Brenner, N. (2013) 'Theses on urbanization', *Public Culture*, 25(1), pp. 85–114.
Brenner, N. (2014) 'Introduction: Urban theory without an outside', in Brenner, N. (ed.) *Implosions/Explosions*. Berlin: Jovis, pp. 14–31.
Brenner, N. and Schmid, C. (2014) 'The "urban age" in question', *International Journal of Urban and Regional Research*, 38(3), pp. 731–755.
Brenner, N. and Schmid, C. (2015) 'Towards a new epistemology of the urban?', *City*, 19(2–3), pp. 151–182.
Crary, J. (2013) *24/7*. Los Angeles: Verso.

Deleuze, G. and Guattari, F. (1987) *A Thousand Plateaus: Capitalism and Schizophrenia*. Minneapolis: University of Minnesota Press.

Derickson, K.D. (2015) 'Urban geography I: Locating urban theory in the "urban age"', *Progress in Human Geography*, 39(5), pp. 647–657.

Dunn, N. (2016) *Dark Matters*. Arlesford: Zero Books.

Edensor, T. (2015) 'The gloomy city: Rethinking the relationship between light and dark', *Urban Studies*, 52(3), pp. 422–438.

Ekrich, A.R. (2006) *At Day's Close*. London: Phoenix.

Guattari, F. (1993) 'Space and corporeity', *Columbia Documents of Architecture and Theory*, 2, pp. 139–149.

Gwiazdzinski, L. (2005) *La Nuit, dernière Frontière de la Ville*. La Tour-d'Aigues: Editions de l'Aube.

Harrison, P. (2009) 'In the absence of practice', *Environment and Planning D: Society and Space*, 27(6), pp. 987–1009.

Hölker, F., Moss, T., Griefahn, B., Kloas, W., Voigt, C.C., Henckel, D., Hänel, A., Kappeler, P.M., Völker, S., Schwope, A., Franke, S., Uhrlandt, D., Fischer, J., Klenke, R., Wolter, C. and Tockner, K. (2009) 'The dark side of light: A transdisciplinary research agenda for light pollution policy', *Ecology and Society*, 15(4), p. 13.

Kearnes, M. and van Dooren, T. (2017) 'Rethinking the final frontier: Cosmo-logics and an ethic of interstellar flourishing', *GeoHumanities*, 3(1), pp. 178–197.

Melbin, M. (1987) *Night as Frontier: Colonizing the World after Dark*. New York: The Free Press.

Merrifield, A. (2012) 'The politics of the encounter and the urbanization of the world', *City*, 16(3), pp. 269–283.

Morrissey, J. (2005) 'Cultural geographies of the contact zone: Gaels, Galls and overlapping territories in late medieval Ireland', *Social and Cultural Geography*, 6(4), pp. 551–566.

Naum, M. (2010) 'Re-emerging frontiers: Postcolonial theory and historical archaeology of the borderlands', *Journal of Archaeological Method and Theory*, 17(2), pp. 101–131.

Oswin, N. (2016) 'Planetary urbanization: A view from outside', *Environment and Planning D: Society and Space*, online early access: http://journals.sagepub.com/doi/abs/10.1177/0263775816675963 (accessed 21 October 2017).

Patel, R. (2006) 'Working the night shift: Gender and the global economy', *ACME: An International e-Journal for Critical Geographies*, 5(1), pp. 9–27.

Pottharst, M. and Könecke, B. (2013) 'The Night and its Loss', in Henckel, D., Thomaier, S., Könecke, B., Zedda, R. and Stabilini, S. (eds) *Space–Time Design of the Public City*. Dordrecht: Springer Netherlands, pp. 37–48.

Pritchard, S.B. (2017) 'The trouble with darkness: NASA's Suomi satellite images of earth at night', *Environmental History*, online early access: https://academic.oup.com/envhis/article-abstract/22/2/312/2998686/The-Trouble-with-Darkness-NASA-s-Suomi-Satellite (accessed 21 October 2017).

Self, W. (2014) 'If you want to see London with completely new eyes, take a night-hike out of town', *New Statesman*, 18 September: www.newstatesman.com/culture/2014/09/will-self-if-you-want-see-london-completely-new-eyes-take-night-hike-out-town (accessed 21 October 2017).

Shaw, R. (2015) 'Night as fragmenting frontier: Understanding the night that remains in an era of 24/7', *Geography Compass*, 9(12), pp. 637–647.

Smith, N. (1984) *Uneven Development*. Oxford: Basil Blackwell.

Smith, R.G. (2013) 'The ordinary city trap', *Environment and Planning A*, 45(10), pp. 2290–2304.
Stone, T. (2017) 'The value of darkness: A moral framework for urban nighttime lighting', *Science and Engineering Ethics*, online early access: https://link.springer.com/article/10.1007/s11948-017-9924-0 (accessed 21 October 2017).
Storper, M. and Scott, A.J. (2016) 'Current debates in urban theory: A critical assessment', *Urban Studies*, 53(6), pp. 1114–1136.
Thomas, D. (1966) *Collected Poems, 1934–1952*. London: Dent.
Valentine, G. (1989) 'The geography of women's fear', *Area*, 21(4), pp. 385–390.

Index

24/7 era 38

abuse 105–6
Adkins, E. 60
aesthetics *see* night-time aesthetics
affects and affectivity 45–6, 47, 73–4, 76, 101
age *see* elderly people; young people
alcohol consumption 41–2, 68–9, 70–2, 73, 74
alcohol industry 70–2, 77
alcohol-related crime 40, 77
Anderson, B. 45
anthropological light pollution 55–6
anti-globalization 114
architectural visualizations 92
art 88–9
artificial lighting
  and atmosphere 76
  as condition of possibility 52–4, 57
  criticism of 57, 58, 114
  development of 29–31, 34, 51, 57–62, 99
  domestic 31, 99, 102
  and energy usage 56
  impact of 51, 52
  light pollution 51, 52, 54–6, 57, 114
  and night-time aesthetics 85–6
  and planetary urbanization 21, 111–12
  uneven access to 33–4, 60
  see also public lighting; light pollution
assemblage 13, 14, 74, 115
astronomical light pollution 54, 57
astronomy 117
atmospheres 35, 75, 76, 78, 79

Baldwin, P.C. 31
Ball, M. 10
banalization 36

Bateson, G. 17
Beaumont, M. 28, 87
Berardi, F. 100
Berlant, L. 43
The BFG 83
Bille, M. 99
biogeoastronomical night 50–1, 112
Blunt, A. 100–1
Bordin, G. 27
Brenner, N. 15–16, 21, 22
'bundling' 103
Buse, C. 101

candles 29–30
capital and urbanization 15, 17–18
capitalism
  and affects of incessancy 43–4
  and domestic sphere 99–100, 107
  night-time production 32–3
  spread into 'everyday life' 15, 37–8
carbon use 57, 59
care of the self activities 102
cities
  changing aspects of 9
  concept and nature of 10–13
  entrepreneurialization of 90–1
  limits to 21–3
  as machines 8, 18–19, 73, 114
  nocturnal subjectivities of 73–5
  representation of *see* night-time aesthetics
  role of night in understanding 9–10
  'three ecologies' in 16–19
  see also urban exploration; urban life; urban night; urbanization
cleaning 62–4
clock-based temporalities 31–2
colonization of night 34, 35–6, 113
Comedia 70

comparative approach 13–14
conditions of possibility 52–4, 57
connectivity 99–100, 102–3, 106
contact zones 45, 80, 94, 115
control 20
  by state 29, 53, 70, 77–8, 86
  and domestic night 98–100
  and night-time economy 70, 77–8
conviviality 86
cosmopolitanism 17
counter-aesthetics of night 87–90
counterculture 69–70
Crary, J. 38, 39–40
creative class 90–1
crime 40, 61, 77
criminals 40–1
cultural representations of night 84–5, 86, 87–9, 119
curfews 28

danger 28, 40, 61, 84, 105–6
Dark Sky Movement 57, 114
Dark Sky Places 57, 112, 117
darkness 2, 28, 84–5, 103–5
Davis, M. 12, 13
de Certeau, M. 17
Deleuze, G. 19, 20
demons 84
Derickson, K.D. 14
'deviants' 40–1
Dickens, C. 41, 87–8
discourses 74
diversity, in research 118
Dogberry 86
domestic activities 39, 102
domestic lighting 31, 60, 99
domestic night
  at edge of urban life 105–7, 117
  and control 98–100
  and fear 97
  and subjectivity 100–5
domestic sphere
  connection with public 102–3
  transference of incessancy 35, 64
domestic violence 105–6
domestic work 39, 102
drinkers 41–2
  see also alcohol consumption
dynamic lighting 59–60

Earth Hour 57
ecological crisis 17, 18
ecological impact of artificial light 51, 52, 55

ecological light pollution 55, 57
ecologies, Guattari's 16–19
Edensor, T. 86
Ekrich, A.R. 27, 28–9, 50, 84, 103
elderly people 101
electric lighting
  development of 30–1, 34, 51, 58–62, 99
  and energy usage 56
  impact of 51, 52
  see also light pollution
electricity (domestic) 31, 99, 102
electrification 30–1, 33, 51
elite 11
  see also creative class
employment sectors 39
empowerment 36
encounter 17
energy usage 56, 57, 59, 102
entrepreneurialization of cities 90–1
environment 17, 18
  impact of artificial light on 51, 52, 55
ethnicity 4, 64
exploration 41, 89–90
explosion 36

fast-food venues 72
fear 28, 40, 61, 84, 97, 105–6
festivals 85
film noir 88
Florida, R.L. 90
Foucault, M. 29, 52–3
fragmentation of frontier 44–5, 94–5, 115–16
frontier
  fragmentation of 44–5, 94–5, 115–16
  night as 34–8, 44–5, 80, 93–5, 115–16

Galinier, J. 27, 28
Garrett, B. 89
gas lighting 30
gender
  and lighting provision 60
  and night-time economy 69, 71
  and night-walking 40, 41
  and shift work 39, 64
Glennie, P. 32
global middle class 90–1
Global North 11, 60–1, 118
Global South 11, 60, 61
global urbanization 16–20, 94, 111–12
Goffman, E. 62
Gothic aesthetics 84–5
Graham, S. 53

Guattari, F. 8–9, 16–19, 73
  see also ecologies, Guattari's
Gwiazdzinski, L. 36, 69

haecceity 3, 109
Hall, S. 11
harmonization 36
health 55, 64
Hobbs, D. 75
Hollands, R. 74
Holmes, Sherlock 86
home
  concepts of 100–3
  see also domestic night
homeless people 42–3
homemaking activities 39, 102
hunting 28
'hygge' 99

incessancy 35, 43–4, 113–14
inequalities 4, 18, 21, 60, 61
infrastructure
  cleaning and maintenance 62–4
  and conditions of possibility 53–4, 57
  developments in lighting 57–62
  home connected to 99, 106
  impact of expanding 51–2
  and subjectivity 74–5
interdisciplinary research 118–19
International Dark Sky Movement 57, 114
intimacy 103, 105
isolation 101–2, 105

Kampala 78–9
karaoke 78–9
Katz, L. 42
Kiribati 7
Könecke, B. 114
Koslofsky, C. 85
Kumar, A. 60, 102, 106

Lamé, A. 92
LED lighting 58–60, 61–2, 65
Lefebvre, H. 15, 37–8
legislation 70, 77–8
LGBT communities 4, 40
light
  and dark 2
  and subjectivity 104–5
light pollution 51, 52, 54–6, 57, 114
light therapy 55
lighting see artificial lighting
lighting festivals 85

literature 84–5, 86, 87–8, 119
Litre of Light project 60
loneliness 101–2, 105
Longcore, T. 55
Lowe, P. 106

McFarlane, C. 13–14
maintenance work 62–4
marketing of place 90–3
marketization and domestic sphere 99
Marx, K. 32, 33, 39
Marxist theory 15
materiality 74–5, 100–1
mechanization of industry 32–3
megacities 12
Melbin, M. 34–5, 63
Merleau-Ponty, M. 103
Merrifield, A. 21, 116
Merz, C. 31
Minkowski, E. 103–4
modernity 12
Møhl, P. 102
Morton, T. 50
mothers 101–2
Much Ado about Nothing 86
myth 28, 50, 84

natural night see biogeoastronomical night
nature 50
Naum, M. 45
neoliberalism 19–20, 99–100
New Year's Eve 7–8
night
  biogeoastronomical night 50–1, 112
  colonization of 34, 35–6, 113
  conceptualizing 2–3, 119
  and fear 28, 40, 61, 84, 97, 105–6
  fragmentation of 44–5, 94–5, 115–16
  as frontier 34–8, 44–5, 80, 93–5, 115–16
  at home see domestic night
  loss of 113–14
  narratives of 112–17
  in pre-industrial societies 27–9
  relationship with darkness 28
  relationship with nature 50
  as research topic 1–4, 117–19
  role in understanding cities 9–10
night markets 76–7
night mayors 42, 92
night sky 54, 57, 117
night workers 35, 38–40, 64
night-time activity

cleaning and maintenance 62–4
  domestic 39, 102
  impact of lighting on 30
  in pre-industrial societies 28–9
  see also night-time economy
night-time aesthetics 83–4, 93–4
  artistic depictions 88–9
  evolution of 84–7
  exploration 89–90
  film noir 88
  night-walkers 87–8
  place marketing 90–3
night-time economy
  and contact zones 80
  diversity of 69
  global examples 70–3, 76–9
  participants 41–2, 71, 74
  portrayal of 72
  in pre-industrial societies 29
  research on 2
  use of term 70
night-time infrastructure
  cleaning and maintenance 62–4
  and conditions of possibility 53–4, 57
  developments in lighting 57–62
  home connected to 99, 106
  impact of expanding 51–2
  and subjectivity 74–5
night-time leisure venues 70–2, 74, 77–8, 91, 92
night-time production 31–3
night-time subjectivities 73–5, 100–5
night-walkers 28, 30, 41, 58, 87–8
nightclubs 72, 78
Nightcrawler 88
nightlife
  and place marketing 91–2
  see also night-time leisure venues
nightology 117–19
nightwatchmen 86
nocturnals 38–43
nyctophobia 26

Oslo 59
Oswin, N. 114–15

performance 84
Pink, S. 101
place marketing 90–3
planetary urbanization 3, 15–16, 110–12
  challenges to 115–17
  and infrastructural expansion 52
  and limits to city 21–3
  and temporal expansion 113–15

plumbing 99
police forces 86
post-globalization 11, 19
postcolonial theory 13–15, 45
Pottharst, M. 114
power 20, 21, 35, 106
  see also state control
practice theory 14
pre-industrial societies 27–9, 32
production at night 31–3
prostitution 79
protection 101
public lighting 29–31, 33–4, 51, 56, 57–62
public sphere, and domestic sphere 102–3
pubs 70–1, 72, 74

refugees 41
regeneration projects 71
representation *see* night-time aesthetics
rest *see* sleep and rest
rhythmanalysis 37–8
Rich, C. 55
ritual 28
road traffic accidents 61
Robinson, J. 12–13, 14
Romeo and Juliet 103

Savage, M. 11
Schivelbusch, W. 85
Schlör, J. 29, 30, 40
Schmid, C. 15–16, 22
seasonal patterns 28
self 16, 104–5
  see also subjectivity/ies
sex work 79
Sharpe, W. 93
shift work 35, 38–40, 64
sleep and rest 2, 27, 38
sleep trackers 99–100
smart lighting 59–60, 61
social class 39, 90–1
  see also elite
social relations 17–18, 100–3
socio-economic inequalities 4, 18, 21, 60, 61
solar-based lighting 60
spectacle (night as) 87, 89–90, 92, 93
spectacular (night as) 85, 89–90, 91, 92, 93
state, home connected to 98–100
state control 29, 53, 70, 77–8, 86
street cleaners 63

street lighting
  at global scale 62
  development of 29, 30, 31, 51, 59–60, 61
  and energy usage 56
street violence 77–8
subjectivity machine 4, 9, 18–20, 73, 114
subjectivity/ies 17, 19
  and domestic night 100–5
  and inequality 4, 21
  of urban night-time 73–5
Swan, J. 30–1
Sydney 77–8

Taipei, night markets in 76–7
task-based temporalities 31–2
technology 57–62, 99–100
temporal expansion 113–15
temporalities, from task to clock 31–2
Thomas, D. 119
Thompson, E.P. 31
'three ecologies' see ecologies, Guattari's
Thrift, N. 32, 53
Time Use Survey 102
time zones 7
Tolia-Kelly, D. 101
Twigg, J. 101

urban 10–11
urban development 71, 92
urban exploration 41, 89–90
urban life
  diverse actants in 20–1
  domestic night at edge of 105–7, 117
  Guattari's understanding of 8–9
  moments beyond 116–17
urban night
  narratives of 112–17
  representation of 83–4
  subjectivities of 73–5
  see also night-time aesthetics; night-time economy; night-time infrastructure
urban theory 10–16, 110–11
  see also planetary urbanization
urbanization
  and capital 15, 17–18
  and post-globalization 11
  see also global urbanization; planetary urbanization

violence 77–8, 105–6
vision 26
vulnerability 105
  see also fear

warmth 56
Wasswa-Matovu, J. 79
wealth 11, 90
Wetherspoons pub chain 74
Wirth, L. 10
workers (nocturnal) 35, 38–40, 63, 64
world cities 12
World Wide Fund for Nature 57

young people 42, 74